# Advances in Experimental Medicine and Biology

Volume 1452

*Advances in Experimental Medicine and Biology* provides a platform for scientific contributions in the main disciplines of the biomedicine and the life sciences. This series publishes thematic volumes on contemporary research in the areas of microbiology, immunology, neurosciences, biochemistry, biomedical engineering, genetics, physiology, and cancer research. Covering emerging topics and techniques in basic and clinical science, it brings together clinicians and researchers from various fields.

*Advances in Experimental Medicine and Biology* has been publishing exceptional works in the field for over 40 years, and is indexed in SCOPUS, Medline (PubMed), EMBASE, Reaxys, EMBiology, the Chemical Abstracts Service (CAS), and Pathway Studio.

2022 CiteScore: 6.2

Heide Schatten
Editor

# Cell and Molecular Biology of Ovarian Cancer

## Updates, Insights and New Frontiers

Springer

*Editor*
Heide Schatten
Department Veterinary Pathobiology
University of Missouri-Columbia
Columbia, MO, USA

ISSN 0065-2598 ISSN 2214-8019 (electronic)
Advances in Experimental Medicine and Biology
ISBN 978-3-031-58310-0 ISBN 978-3-031-58311-7 (eBook)
https://doi.org/10.1007/978-3-031-58311-7



This Springer imprint is published by the registered company Springer Nature Switzerland AG
The registered company address is: Gewerbestrasse 11, 6330 Cham, Switzerland

# Preface

Ovarian cancer is the second most common gynecological cancer in women, ranging third of the most common cancers among females after breast and lung cancer. Ovarian cancer still has the highest mortality rate being the fifth leading cause of cancer death among women in the United States.

The high mortality rate is in part related to the difficulties to diagnose the disease in early stages, as reliable ovarian tumor biomarkers are still not yet available, and patients do not experience symptoms at early stages of disease development. Late-stage ovarian cancers display various heterogeneous groups of histological features which makes treatment options more difficult and requires more treatment modalities and extensive surgical cytoreduction. Because of the severity of the disease in late stages, it has become an urgent research priority to find new biomarkers and new general health markers to diagnose the disease in early stages and determine effective and specific targeted treatment strategies.

So far, standard treatment for ovarian cancers includes surgery and platinum-based chemotherapy and treatment with taxanes. New clinical trials include the evaluation of new targeted therapeutic agents and immunotherapy combined with primary therapies to increase the survival rate in patients. In addition, extensive screening and research in animal models is aimed at early detection to design new strategies to reduce the currently high mortality rate of ovarian cancer patients and improve the overall 5-year survival rate that remains low at 40% for stage III and 20% for stage IV.

Encouraging new research has been initiated on several levels to combat the disease which in part is based on extensive basic research on cell and molecular biology levels as well as on modification and screening of existing and new drugs to determine more efficient targeted chemotherapies. There are still many unknown factors leading to manifestation of the disease which are currently being addressed in several excellent research labs. New biomarkers are being determined based on the analysis of abnormal signaling events in cancer cells with the goal to repair abnormalities on cellular levels and restore normal cellular functions. It is further being explored why the relapse rate after treatment is still high with relapse occurring in 65–75% of patients. Drug resistance is among the various aspects being investigated to improve currently administered drugs. New approaches also include analyses of general health aspects with new research on the role of the microbiome in disease which is a rapidly evolving highly promising emerging field in the health sciences.

As also indicated in the preface of the companion book on molecular and diagnostic imaging and treatment strategies of ovarian cancer, the advent of molecular and genomic technologies has significantly improved our understanding of the biological processes underlying ovarian cancer which has been enhanced by the Cancer Genome Atlas that has identified mutations in human ovarian cancer genomes that may play a role in tumor progression and modifications of cellular metabolism. Targeted therapies are now available to inhibit specific signaling pathways that are aberrant in ovarian cancer cell populations, and we are now able to image signaling molecules with specific markers in live cells in culture. Progress has also been made in designing nanoparticles that offer the potential for imaging and targeted ovarian cancer treatment. The joint initiatives and efforts of advocate patients, ovarian cancer survivors, basic researchers, statisticians, epidemiologists, and clinicians with various and specific expertise have allowed close communication for more specific and targeted treatment. Major forces supporting these efforts are the Department of Defense, the American Cancer Society, and several other Foundations that recognized the need for intensified advocacy to find treatments for the disease that represents an under-studied area of research. Multi-modal approaches are oftentimes required to manage ovarian cancer and achieve positive outcomes which require patient-specific evaluation and analysis for specific management.

The present book on cell and molecular biology of ovarian cancer is the second one of two companion books, with the second one being focused on specific aspects of cell and molecular biology of ovarian cancer. Both books include new and exciting aspects of ovarian cancer research with chapters written by experts in their respective fields who contributed their unique expertise in specific ovarian cancer research areas and include cell and molecular details that are important for the specific subtopics. Comprehensive and concise reviews are included of key topics in the field. Cutting-edge new information is balanced with background information that is readily understandable for the newcomer, ovarian cancer patients, and for the experienced ovarian cancer researcher alike. Chapters include microtubule-targeting agents: disruption of the cellular cytoskeleton as a backbone of ovarian cancer therapy; tubulin complexity in cancer and metastasis; the impact of centrosome pathologies on ovarian cancer development and progression with a focus on centrosomes as therapeutic target; drug-resistant epithelial ovarian cancer: current and future perspectives; the role of the human microbiome in epithelial ovarian cancer; insights into the microbial composition of intratumoral, reproductive tract, and gut microbiota in ovarian cancer patients; and the impact of mitochondria in ovarian cancer cell metabolism, proliferation, and metastasis.

I hope that this book will stimulate further advances in ovarian cancer research and contribute new insight into potential new targets for ovarian cancer therapies. I am most grateful and would like to express my sincere thanks to the publisher for inviting this book and the companion book on molecular and diagnostic imaging and treatment strategies of ovarian cancer with special thanks to Tiffany Lu and associates for all their excellent help and care during all stages of the project.

It is a great pleasure and timely to edit this book on cell and molecular biology of ovarian cancer depicting areas in ovarian cancer that have impacted new treatment strategies. I am most grateful to the outstanding contributors for sharing their unique and specific expertise with the scientific community. My sincere thanks to all for their most valuable contributions.

Columbia, MO, USA Heide Schatten

# Contents

**1 Microtubule-Targeting Agents: Disruption of the Cellular Cytoskeleton as a Backbone of Ovarian Cancer Therapy** ..... 1
Michael Danziger, Helen Noble, Dana M. Roque, Fuhua Xu, Gautam G. Rao, and Alessandro D. Santin

**2 Tubulin Complexity in Cancer and Metastasis** .............. 21
Michael Danziger, Fuhua Xu, Helen Noble, Peixin Yang, and Dana M. Roque

**3 The Impact of Centrosome Pathologies on Ovarian Cancer Development and Progression with a Focus on Centrosomes as Therapeutic Target** .................................. 37
Heide Schatten

**4 Drug-Resistant Epithelial Ovarian Cancer: Current and Future Perspectives** ................................ 65
Megha Mehrotra, Pratham Phadte, Priti Shenoy, Sourav Chakraborty, Sudeep Gupta, and Pritha Ray

**5 The Role of the Human Microbiome in Epithelial Ovarian Cancer** ........................................ 97
Diane Mahoney

**6 Insights into the Microbial Composition of Intratumoral, Reproductive Tract, and Gut Microbiota in Ovarian Cancer Patients** .................................... 107
Qian Zhou and Qingren Meng

**7 The Impact of Mitochondria in Ovarian Cancer Cell Metabolism, Proliferation, and Metastasis** .................. 119
Heide Schatten

**Index** .................................................... 127

# 1 Microtubule-Targeting Agents: Disruption of the Cellular Cytoskeleton as a Backbone of Ovarian Cancer Therapy

Michael Danziger, Helen Noble, Dana M. Roque, Fuhua Xu, Gautam G. Rao, and Alessandro D. Santin

**Abstract**

Microtubules are dynamic polymers composed of α- and β-tubulin heterodimers. Microtubules are universally conserved among eukaryotes and participate in nearly every cellular process, including intracellular trafficking, replication, polarity, cytoskeletal shape, and motility. Due to their fundamental role in mitosis, they represent a classic target of anti-cancer therapy. Microtubule-stabilizing agents currently constitute a component of the most effective regimens for ovarian cancer therapy in both primary and recurrent settings. Unfortunately, the development of resistance continues to present a therapeutic challenge. An understanding of the underlying mechanisms of resistance to microtubule-active agents may facilitate the development of novel and improved approaches to this disease.

**Keywords**

Microtubule · Tubulin · Dynamic instability · Paclitaxel · Docetaxel · Ixabepilone · Vincristine · Vinblastine · ABC transporter

## 1.1 Microtubules

Microtubules are cylindrical filaments consisting of α- and β-tubulin subunits which remain universally conserved among eukaryotes and participate in nearly every cellular process, including replication, organelle trafficking, cytoskeletal structure, cell polarity, and motility. Even many prokaryotes have *ftsZ* or similar genes encoding a tubulin homolog that assembles into protein polymers involved in cellular division (Goodson and Jonasson 2018). As such, microtubules are one of the most basic and essential elements of the eukaryotic cell. Mutations resulting in derangement in normal microtubule function contribute to a panoply of human disease, the

M. Danziger · H. Noble
Department of Obstetrics, Gynecology, and Reproductive Sciences, University of Maryland School of Medicine, Baltimore, MD, USA

D. M. Roque · F. Xu · G. G. Rao
Division of Gynecologic Oncology, Greenebaum Comprehensive Cancer Center, University of Maryland School of Medicine, Baltimore, MD, USA

A. D. Santin (✉)
Yale School of Medicine, Yale Cancer Center, New Haven, CT, USA
e-mail: alessandro.santin@yale.edu



H. Schatten (ed.), *Cell and Molecular Biology of Ovarian Cancer*, Advances in Experimental Medicine and Biology 1452, https://doi.org/10.1007/978-3-031-58311-7_1

best characterized of which include neurodegenerative disorders such as Alzheimer's disease (Robert and Mathuranath 2007). Given the role of microtubules and their associated proteins in the cell cycle and cytokinesis, it remains unsurprising that dysregulation of microtubule function occurs in many cancers in the setting of constitutive upregulation of proliferation with consequent rapid cell division (Chandrasekaran et al. 2015; Dráber and Dráberová 2021; Hanahan and Weinberg 2011). As a result, antitumor strategies targeting microtubules are used in the treatment of a number of gynecologic cancers, including ovarian, cervical, and endometrial.

### 1.1.1 Formation and Features

Microtubules are dynamic heterodimeric polymers of α- and β-tubulin. In humans, there are at least nine isoforms of α- and β- tubulin each. Microtubule functional diversity, the "tubulin code," is governed not only by differential expression of the isoforms but also by post-translational modifications (reviewed in the Chap. 2) (Ferreira et al. 2018). Their structure and dynamism are essential to cellular functions, including cytoskeletal organization, mitosis, cell motility, vesicular and organelle transport and signaling, among others. The αβ-tubulin heterodimer pairs are organized head-to-tail (i.e., such that the "head" of a β-tubulin subunit always interfaces with the α-tubulin "tail" of the adjacent αβ-tubulin heterodimer) into 13 protofilaments, which form a hollow polar tube with all heterodimers pointing in the same direction. Lateral interactions between staggered neighboring protofilaments that result in homotypic α-α and β-β contacts provide internal stability of the tube, producing a helical lattice 25 nm in diameter. This internal stability is critical for the dynamic activity characteristic of microtubules because it means that polymerization and depolymerization can only occur at the ends of the tube. The tubes have a plus end, characterized by exposed β-tubulin, and a minus end, characterized by exposed α-tubulin. Heterodimers are added most rapidly to the plus end (Conde and Cáceres 2009; Akhmanova and Steinmetz 2015; Brouhard and Rice 2018). This polarity gives rise to two distinctive microtubule behaviors: dynamic instability and treadmilling. Interestingly, both occur at significantly slower rates in vitro compared to in vivo, highlighting the essential role of microtubule-associated proteins (MAPs) in directing and optimizing the processes (Rodionov et al. 1999; Vasquez et al. 1994).

***Dynamic instability*** Dynamic instability is a stochastic process by which the ends of microtubules rapidly shift between states of growth and shrinkage. Dynamic instability is a form of non-equilibrium polymerization that is facilitated by the hydrolysis of guanosine triphosphate (GTP). Each α- and β-tubulin subunit has a GTP binding site, though only the β subunits hydrolyze GTP. The β-tubulin subunit site to which GTP binds is the E (exchangeable) site. By contrast, the α-subunit binds GTP at the N (non-exchangeable) site where it serves a structural role but is not hydrolyzed (Desai and Mitchison 1997; de Forges et al. 2012; Zhang et al. 2015). When αβ-tubulin heterodimers bind to the plus end of a microtubule, the β-tubulin subunit hydrolyzes GTP to guanosine diphosphate (GDP). As the plus end of the microtubule grows, there will be a considerable portion of the microtubule in which the β-tubulin subunits are bound to GDP. β-tubulin subunits bound to GDP are more prone to depolymerization because of conformational changes to the α-tubulin following GTP hydrolysis of the β-tubulin subunit that place strain on the lattice. If the number of αβ-tubulin heterodimers exceeds the critical concentration ($C_c$, the amount of monomer necessary to achieve polymerization), a new subunit will be added to the plus end before the previously added subunit is able to hydrolyze GTP. This results in a persistent state of GTP-bound β-tubulin at the plus end. This state, in which an abundance of free subunits permits continual addition to the plus end, is referred to as the *GTP cap* (Cassimeris and Spittle 2001; Brouhard and Rice 2018). The GTP cap thus prevents depolymerization. When the rate of hydrolysis equals or surpasses the rate at which new subunits are added, the result is rapid depolymerization of the microtubule, a process called *catastrophe*. Microscopically, protofilaments of microtubules undergoing catastrophe appear to peel away from the lattice, a behavior that is unsurprising given that microtubules are often observed to have curvature in vivo. Catastrophe is reversed by the process of *rescue*, in which the GTP cap is regained by the addition of new subunits (Desai and Mitchison 1997; Goodson and Jonasson 2018; Brouhard and Rice 2018).

***Treadmilling*** Microtubule growth occurs at both the plus and minus ends, though it does so faster at the plus end. The plus end of the microtubule is more prone to catastrophe than the minus end, while the minus end has a higher rate of rescue than the plus end. Consequently, the predominant activity of microtubules (i.e., growth vs. catastrophe) occurs at the plus end. Treadmilling occurs when the subunits are added to the plus end and lost from the minus end at the same rate. Treadmilling thus represents a steady state for a given microtubule. Microtubules may display primarily dynamic instability, treadmilling behavior, or a mixture of the two (Margolis and Wilson 1981; Rodionov and Borisy 1997; Gadde and Heald 2004; Ganem and Compton 2006).

Treadmilling is a facet of protein polymer behaviors that was first described in actin filaments by Wegner (1976). Microtubule treadmilling has subsequently been demonstrated both in vitro and in vivo (Margolis and Wilson 1981; Chen and Zhang 2004). Additional work has demonstrated that treadmilling is a distinct behavior from dynamic instability and that the two are, collectively, aspects of microtubule dynamics that are essential to broaden cellular functionality of microtubules. Treadmilling of kinetochore microtubules is more specifically called "fluxing" or "poleward flux." This process involves the continual translocation of spindle microtubules poleward coupled to disassembly at the minus ends. Loss of tubulin from the minus ends during metaphase is balanced by treadmilling of units to the plus ends, bound to the kinetochore. The net effect is to maintain constant spindle length while tubulin subunits are constantly flowing poleward (Ganem and Compton 2006; Buster et al. 2007).

### 1.1.2 Role in Cellular Function

Microtubules, either directly or indirectly, participate in almost every cellular function, including the cell cycle, cytokinesis, intracellular organization, trafficking of organelles, as well as cellular motility and polarity. Their dynamic nature allows them to adapt in response to the changing needs of the cellular environment. Microtubule-associated proteins (MAPs) further influence and regulate microtubule functionality. Mutations affecting microtubules and their functions occur often at the level of MAPs (reviewed in Chap. 2).

***Mitosis and Chromosome Segregation*** One of the most recognized functions of microtubules is in organization of the mitotic spindle (McIntosh 2016). Numerous changes in microtubules must occur as the cell progresses from a resting state (i.e., interphase) to active replication (i.e., mitosis). Interphase microtubules begin to disappear as protein synthesis slows and are replaced by microtubules nucleating from centrosomes. These microtubules are more labile due to phosphorylation of MAPs that regulate dynamics, including stathmin. As the interphase microtubules disappear, the cell becomes more rotund, granting it a symmetry that will be useful later in cytokinesis. During this transition, the centrosome duplicates to produce a bipolar mitotic spindle. As the new microtubules grow, the mitotic spindle forms. Three species of microtubules comprise the mitotic spindle: kinetochore microtubules, astral microtubules, and non-kinetochore (or interpolar) microtubules. Kinetochore microtubules nucleate from the centrosome and the plus ends bind the kinetochore attached to each sister chromatid. The kinetochore is a large, disk-shaped multiprotein structure that attaches to an area of condensed satellite DNA on the chromatid called the centromere. Each kinetochore binds multiple cross-linked microtubules called K-fibers that serve as powerful anchors during separation of the daughter cells (anaphase). Astral microtubules radiate from the centrosome but do not bind the kinetochore and serve to help position the mitotic spindle within the cell itself. Non-kinetochore microtubules are densely packed between the poles and crosslink with associated proteins to perform a structural scaffold for the spindle. Motor proteins belonging to the family of kinesins and dyneins bind astral and non-kinetochore microtubules and generate polarity. Kinesins engage in the mitotic spindle. For instance, the tetrameric protein kinesin-5 contains two dimeric motor domains at each end that can be oriented in opposite directions to bind antiparallel microtubules at their respective plus ends and slide them in opposite directions. The result of this is to push the minus ends of the microtubules polewards. Dynein serves two functions in the mitotic spindle: (1) attaching to the minus end of a microtubule formed in the body of the spindle and transporting it to the minus end of one nucleated at the centrosome, resulting in elongation of microtubules nucleated at the centrosome and (2) binding the plus end of astral microtubules and pulling them to the cell cortex, maintaining spindle polarity (Goulet and Moores 2013; McIntosh 2016; Alberts et al. 2017). Following spindle attachment, the aligned chromosomes oscillate under tension due to growth and shortening of the kinetochore microtubules (Shelby et al. 1996). This is coupled with microtubule treadmilling or flux behavior from the poles to the kinetochore during anaphase, which results in the poleward migration of sister chromatids (Mitchison 1989; Chen and Zhang 2004; Buster et al. 2007).

***Intracellular Organization*** The centrosome serves as the microtubule-organizing center (MTOC) of an animal cell. Of note, centrosomes are absent in many fungi and seed plants, many classes of protists, and even mammalian female oocytes (Carvalho-Santos et al. 2011; Courtois et al. 2012). Centrosomes consist of two centrioles, pericentriolar material, and γ-tubulin ring complexes. Centrioles are comprised of nine triplets of specialized microtubules that recruit pericentriolar material. The two centrioles are organized perpendicularly. Pericentriolar material is a matrix that provides structure to the centrosome and contains proteins essential for microtubule nucleation, including γ-tubulin and the accessory proteins that make up the γ-tubulin ring complex (γ-TuRC). The centrosome displays enrichment of γ-tubulin, a specialized tubulin complexed to accessory proteins that helps nucleate microtubules from the MTOC. A centrosome can recruit around 50 copies of γ-TuRC. γ-TuRCs serve as points of nucleation for any microtubule originating from the MTOC; however, because the majority of microtubules are not attached to the MTOC, the cytoplasm contains most of the γ-TuRC in a given cell. Following nucleation, the minus end of the microtubule remains at the centrosome while the plus end grows toward the cell periphery (Akhmanova and Steinmetz 2015; Gadde and Heald 2004; de Forges et al. 2012; Goodson and Jonasson 2018; Kollman et al. 2011). This gives microtubules both an intrinsic negative-to-positive polarity and also a central-to-distal cellular spatial polarity in which the minus ends are located centrally while the plus ends are at the cell periphery. This subsequently confers directionality for intracellular trafficking. This is important because, as discussed below, trafficking in either direction requires motor proteins specific to that polarity.

***Organelle trafficking: kinesin/dynein motors*** Microtubules have two associated motor protein families, kinesin and dynein, that assist with intracellular transport of organelles in opposing directions. Microtubules serve as dynamic tracks along which these motor proteins traffic cargo. Kinesins were the first motor proteins discovered that could move cargo along microtubules (Vale et al. 1985). Since then, the kinesin family has been expanded to include at least 45 distinct kinesin-related proteins in humans. Kinesins share significant sequence homology in their motor domains; however, their unifying function is not transportation of cargo, but rather regulation of microtubule dynamics. Dimeric, hetero-trimeric, and tetrameric kinesins have been characterized. Variation of the kinesin tail domain allows for targeting to different intracellular cargos (Sweeney and Holzbaur 2018). All kinesins that possess a motor domain at the N-terminus are characterized by plus end motor activity while those with motor domains at the C-terminus are characterized by minus end motor activity (Marx et al. 2009). Kinesin-1, the first kinesin identified, is a plus end motor that has two heavy chains per motor domain. These are held together by a coiled-coil tail that helps to mediate heavy-chain dimerization. Kinesin-1 moves along a single protofilament of a given microtubule in a coordinated alternating stepping of the two motor domains driven by adenosine triphosphate (ATP) hydrolysis, which allows for 8-nm "steps" along the microtubule (Schnitzer and Block 1997). The kinesin-1 tail domains serve both cargo binding and autoregulatory functions. First, kinesin-1 transportation of diverse organelle and vesicular cargo throughout the cell (including mitochondria, lysosomes, and endosomes) is mediated by the tail domain of the kinesin-1 heavy chain and the associated light chains (Fu and Holzbaur 2014; Sweeney and Holzbaur 2018). Second, the tail domains can fold, allowing interaction of an IAK motif between the two heads that inactivates the motor. Subsequent liberation of the motif, by contrast, results in an active state (Kaan et al. 2011).

Kinesin function is heterogeneous. Kinesin-5 is a tetrameter involved in mitotic spindle formation, where it binds antiparallel microtubules to provide spindle polarity. Kinesin-8 family proteins promote microtubule depolymerization by localizing to the plus end and blocking translocation of additional tubulin subunits to the microtubule (Varga et al. 2006; Walczak et al. 2010). Kinesin-13 family proteins (which includes mitotic centromere-associated kinesin [MCAK]) lack motor function and instead promote catastrophe at both plus and minus ends (Domnitz et al. 2012; Walczak et al. 2010). Kinesin family proteins have been identified as protooncogenes (e.g., 2C) (Zhang et al. 2022), with expression patterns that correlate with prognosis (Wu et al. 2022), generating interest in the development of kinesin spindle protein inhibitors (Jiang and You 2013).

Dynein is a large motor protein that is best characterized by its minus end motor protein function. Like kinesins, the motor domain of dynein lies within the large heavy chain, though the structural similarities with kinesin end there. Dynein is characterized by a microtubule-binding domain attached by a stalk region to a C-terminal hexameric AAA (*A*TPases *A*ssociated with diverse cellular *A*ctivity) ring (Carter et al. 2011). The first AAA domain is the primary site of ATP binding and hydrolysis; nucleotide binding at other domains has an allosteric impact on dynein activity (Gibbons et al. 1987; Kon et al. 2004). ATP hydrolysis results in conformational shifts that are transmitted to the microtubule binding site. A linker domain connects the ATP-hydrolyzing AAA domain to the tail and is essential for generation of force (Alberts et al. 2017; Roberts et al. 2013, 2012). Dynein subsequently undergoes an ATP-powered stroke (similar to that of the actin-myosin ATPase) that mediates its progress along the microtubule (Holzbaur and Johnson 1989; Sweeney and Holzbaur 2018).

In stark contrast to the structural and functional diversity of the kinesin family, there is only one cytoplasmic dynein that serves as the major minus end-directed motor protein. Like the array of kinesins, cytoplasmic dynein carries out a multitude of intracellular transportation functions. As mentioned earlier, it also serves key functions in the mitotic spindle by both assisting elongation of microtubules from the centromere and pulling astral microtubules toward the cortex to maintain spindle polarity. Dynein-based locomotion may promote metastasis (Tagay et al. 2023).

## 1.2 NCCN-Approved Agents for Use in Ovarian Cancer

Given such an essential role in core cellular processes including division, the microtubule has long been the target of anti-cancer therapies. Both hyper-stabilizing and destabilizing agents are widely employed to inhibit ovarian cancer. All pharmacologically relevant agents currently bind β-tubulin. Reviewed below are agents that have been endorsed by the National Comprehensive

Cancer Network (NCCN) (National Comprehensive Cancer Network 2023).

## 1.3 Microtubule-Stabilizing Agents

### 1.3.1 Paclitaxel

***Characteristics*** One of the most commonly used drugs in oncology, paclitaxel, is a diterpene natural product of the Pacific yew (*Taxus brevifolia*) bark that was the first microtubule-stabilizing agent to be discovered (Wani et al. 1971; Schiff et al. 1979). The US Food and Drug Administration (FDA) approved Taxol® for the treatment of ovarian cancer in 1992. It acts as a microtubule-stabilizer by reversibly binding tubulin to promote microtubule assembly and prevent depolymerization. The latter mechanism in particular prevents mitotic spindle function and, consequently, cellular division (Rowinsky and Donehower 1995). Paclitaxel binds to a site on the β-tubulin luminal side, aptly named the taxoid binding site (Field et al. 2013).

Due to its poor water solubility, paclitaxel is suspended in Cremophor® (BASF, Ludwigshafen, Germany), a non-ionic solubilizer prepared by reacting ethylene oxide with castor oil. In the absence of premedication, Cremophor® contributes to hypersensitivity reactions in 30% of cases (Hájek et al. 1996; Roselló et al. 2017), which are sometimes severe (4%) or even life-threatening (2%) (Tymon-Rosario et al. 2021). Due to the risk of hypersensitivity, patients receiving paclitaxel often receive premedication with dexamethasone, diphenhydramine, and famotidine (Markman et al. 2000). Premedication schemes vary between institutions. Chronic steroid administration becomes problematic for diabetic patients and with weekly paclitaxel administration. Interestingly, a retrospective review of 3181 patients failed to identify any correlation between steroid dosing or route, thereby leading some to propose a single 10 mg dose as optimal (Lansinger et al. 2021). In this study, the rate of hypersensitivity reaction was 8.3% and multivariate analyses showed an increased risk of hypersensitivity for female gender (HR 1.26) and gynecologic cancer (HR 1.34). In breast cancer, prospective data suggests that steroid premedication can be omitted from cycle 3 and 4 of a dose-dense regimen if no hypersensitivity reactions occur during preceding cycles (Barroso-Sousa et al. 2021). The drug is commercially available as an injection concentrate in 30 mg (5 mL), 100 mg (16.7 mL), and 300 mg (50 mL) vials. It must be diluted in solution (normal saline, 5% dextrose, 5% dextrose in normal saline or lactated Ringer's) to a final concentration of 0.3–1.2 mg/mL. Intravenous (IV) administration should include a 0.22 micron filter to remove particulates.

***Efficacy-epithelial histologies*** The conventional scheduling of paclitaxel therapy (in combination with a platinum agent) for epithelial ovarian cancer is an intravenous infusion as adjuvant therapy following debulking surgery for a total of 6 cycles, though weekly dosing and neoadjuvant approaches remain appropriate in certain instances. The FDA-approved regimen (Food and Drug Administration 2011) for previously untreated patients is 175 mg/m$^2$ over 3 h every 3 weeks, or historically 135 mg/m$^2$ over 24 h every 3 weeks (both in combination with platinum). In general, 24-h paclitaxel infusions have been associated with equal efficacy but greater marrow toxicity, inconvenience, and cost; the only benefit of a slower infusion appears to be a reduction in neuropathy (Eisenhauer et al. 1994). Weekly regimens commonly consist of 80 mg/m$^2$ over 1 h on days 1, 8, and 15 every 3 weeks in combination with carboplatin at an area under the curve (AUC) of 5–6 mg/mL/min on day 1 or AUC of 2 mg/mL/min weekly for a total of 6 cycles (Clamp et al. 2019). Neoadjuvant regimens employ 175 mg/m$^2$ over 3 h on day 1 every 3 weeks in combination with carboplatin ideally for 3 cycles, followed by debulking surgery, then an additional 3 cycles. The EORTC 55971 and CHORUS trials defined a role for neoadjuvant chemotherapy in patients with high tumor burden and/or poor performance status to achieve similar survival outcomes with less surgical morbidity compared to primary debulking surgery (Vergote et al. 2018; Kehoe et al. 2015; Vergote et al. 2010; Fagotti et al. 2016).

The importance of dose density appears to vary by population. Paclitaxel administered in weekly fashion putatively exhibits anti-angiogenic effects. In 2013, the JGOG 3016 study showed a benefit in progression-free (PFS) (28.2 vs 17.5 months, HR 0.76, 95% CI 0.62–0.91, $p$ = 0·0037) and overall (100.5 vs 62.2 months, HR 0.79, 95% CI 0.63–0.99, $p$ = 0·039) survival (OS) in women with stage II-IV previously untreated ovarian cancer randomized to receive dose-dense (carboplatin AUC 6 mg/mL/min on day 1 and paclitaxel 80 mg/m$^2$ on days 1, 8, and 15) compared to conventional (carboplatin AUC 6 mg/mL/min and paclitaxel 180 mg/m$^2$ on day 1) treatment within a 21-day cycle (Katsumata et al. 2009). The survival advantages came at the price of worse neuropathy and anemia. Subsequent attempts to recapitulate this in a North American (GOG 262) (Chan et al. 2016) or European (ICON 8) (Clamp et al. 2019) population generally failed, though there were differences in design including allowance of bevacizumab in some studies (Chan et al. 2016). MITO-7 illustrated that newly diagnosed women with stage IC-IV ovarian cancer can achieve equivalent oncologic outcomes with better quality of life using weekly carboplatin AUC 2 mg/mL/min and combined with weekly paclitaxel (MITO 7) (Pignata et al. 2014). An analysis of the Cancer Genome Atlas (TCGA) suggests that mesenchymal (as opposed to immunoreactive, differentiated, proliferative) molecular subtypes may respond particularly well to weekly paclitaxel (Murakami et al. 2016). Differences in pharmacokinetics and pharmacogenomics among ethnic groups may contribute to differential responses and toxicities (Súarez-Zaizar et al. 2020; Johnson et al. 2023).

Intraperitoneal (IP) dosing consists of 60 mg/m$^2$ on day 8 of a 21-day treatment cycle for 6 cycles in combination with IV paclitaxel 135 mg/m$^2$ over 24 h on day 1 and IP cisplatin 100 mg/m$^2$ on day 2 (Armstrong et al. 2006). GOG 172 demonstrated benefit in PFS (23.8 vs 18.3 months) and overall survival (65.6 vs 49.7 months); however, grade 3/4 toxicities were more common in the IP group and only 42% of the IP group completed 6 cycles. Modern dosing schemes suggest that IV approaches that incorporate bevacizumab may be better tolerated with similar survival outcomes (Walker et al. 2019).

Prior to the introduction of taxanes, platinum therapy was administered with cyclophosphamide. GOG 111 demonstrated that paclitaxel 135 mg/m$^2$ over 24 h followed by cisplatin 75 mg/m$^2$ every 3 weeks was superior to cisplatin and cyclophosphamide in patients with advanced epithelial ovarian cancer, with an improved objective response rate (ORR) (73% vs 60%) and improved PFS (18 vs 13 months) and OS (38 vs 24 months) (McGuire et al. 1996). Paclitaxel is not recommended as a first-line single-agent chemotherapeutic based on GOG 132, which showed superior ORR and PFS with cisplatin alone or in combination with paclitaxel compared to paclitaxel alone (though overall survival was similar in all arms) (Muggia et al. 2000).

Initial treatment with carboplatin and paclitaxel leads to remission in 60–80% of patients; however, 85% of patients with advanced-stage disease recur (Rubin et al. 1999). Platinum-free interval (PFI) is defined as the amount of time between completion of platinum-based therapy to disease relapse. A PFI of ≥6 months defines platinum-sensitive disease, while a PFI of <6 months defines platinum-resistant disease. ICON4 demonstrated that the addition of paclitaxel to platinum-based therapy in patients with recurrent platinum-sensitive epithelial ovarian cancer increased PFS (50% vs 40%) and OS (57% vs 50%) (Parmar et al. 2003). Paclitaxel may also be an effective single-agent therapy in patients with platinum-resistant disease, particularly if administered weekly (Gynecologic Oncology Group et al. 2006). The combination of IV paclitaxel weekly at 80 mg/m$^2$ with biweekly bevacizumab 10 mg/kg of a 28-day cycle has shown to be one of the most effective regimens for platinum-resistant disease (Poveda et al. 2015) and superior to weekly paclitaxel alone (Pujade-Lauraine et al. 2014).

***Efficacy-non-epithelial histologies*** Paclitaxel is also a component of the preferred regimens for non-epithelial histologies that affect pre-menopausal women. For recurrent malignant germ cell tumors, paclitaxel can be used at 250 mg/m$^2$ over 24 h on day 1 of a 3-weekly treatment cycle in combination with ifosfamide 1,500 mg/m$^2$ administered by infusion over 60 min daily on days 2 to 5, mesna 500 mg/m$^2$ administered before ifosfamide and at 4 and 8 h after ifosfamide daily, and cisplatin 25 mg/m$^2$ administered by infusion over 30 min daily on days 2 to 5 (TIP), as well as a chemotherapy partner for other palliative doublets (National Comprehensive Cancer Network 2023). These data demonstrate that TIP resulted in a 70% complete response (Kondagunta et al. 2005). Paclitaxel plus gemcitabine has additionally been studied in a phase II trial that showed an objective response rate of 21.4% with 3 complete responders, 2 of whom were disease free at 15 and 25 months (Hinton et al. 2002). Paclitaxel in combination with carboplatin (Brown et al. 2004, 2005) or etoposide and platinum ± bleomycin (BEP) is a common first-line regimen for malignant sex cord stromal tumors. GOG 264, a phase II randomized trial of paclitaxel and carboplatin versus BEP for newly diagnosed advanced stage and recurrent chemonaive sex cord stromal tumors of the ovary, closed to accrual at the interim futility analysis after paclitaxel/carboplatin showed a favorable toxicity profile but no improvement in PFS relative to BEP (HR 1.12, 95% CI 0.58–2.16); median PFS was 27.7 months for PC and 19.7 months for BEP (NRG Winter Meeting (virtual), 2021). As malignant germ cell and sex cord stromal tumors constitute only 5% and 7% of all ovarian cancers, respectively (Gershenson et al. 2019), their rarity complicates the completion of randomized controlled trials.

***Toxicity*** Neutropenia and neuropathy pose dose-limiting toxicities. Hematologic nadir occurs at a median of 11 days and is associated with higher doses and longer duration of infusion (Marupudi et al. 2007). Sensory neuropathy may present as a mixture of paresthesias and dysesthesias in a stocking-glove distribution (Kuroi and Shimozuma 2004). Alterations in axonal transport and endocytosis (Shin 2023) or metabolic variations (Johnson et al. 2023) may underlie predisposition to paclitaxel-induced neuropathy. Other toxicities include EKG abnormality, alopecia, diarrhea, nausea and vomiting, stomatitis, and myalgias and arthralgias.

### 1.3.2 Nab-paclitaxel

***Characteristics and efficacy*** Nanoparticle albumin-bound paclitaxel (nab-paclitaxel) mitigates infusion reactions and neurotoxicity. Due to differences in volume of distribution and other pharmacokinetics, nab-paclitaxel may result in favorable tissue penetration and slower elimination than solvent-bound paclitaxel (Chen et al. 2014). The phase II GOG 126R study examined the efficacy and toxicity of nab-paclitaxel as a single-agent therapy for the treatment of platinum- and taxane-resistant ovarian cancer and found an ORR of 23%, PFS of 4.5 months, and OS of 17.4 months. Additionally, no grade 4 or 5 toxicities were observed (Coleman et al. 2011). A phase II study demonstrated an objective response rate of 64% in patients with recurrent platinum-sensitive disease (Teneriello et al. 2009). As a result, it may serve as a reasonable replacement in patients who are unable to tolerate conventional Paclitaxel therapy (Maurer et al. 2017).

***Dosing*** As a single agent, a recommended dosage of nab-paclitaxel is 260 mg/m$^2$ IV on day 1 of a 21-day cycle for 6–8 cycles or 100 mg/m$^2$ on days 1, 8, and 15 of a 28-day cycle until disease progression or unacceptable toxicity (Coleman et al. 2011; Teneriello et al. 2009). No premedication is necessary with nab-paclitaxel, in contrast to its parent formulation. Mauer and colleagues (2017) reported on 37 patients for whom nab-paclitaxel was used in patients with solvent-based paclitaxel hypersensitivity weekly at 60–100 mg/m$^2$ ($n$ = 12) or three-weekly at 135 mg/m$^2$ ($n$ = 13), 175 mg/m$^2$ ($n$ = 9), or 225 mg/m$^2$ ($n$ = 3); thirty four patients (91.9%) received nab-paclitaxel in combination with either carboplatin ($n$ = 28, 75.7%), IP cisplatin (*n* = 1, 2.7%), carboplatin and bevacizumab (*n* = 3, 8.1%), or carboplatin and gemcitabine (*n* = 2, 5.4%) (Maurer et al. 2017).

***Toxicity*** Myelosuppression, in the form of neutropenia, is the main hematologic toxicity. While the rate of hypersensitivity reactions appears to be less than that of standard paclitaxel, these reactions may still occur. Sensory neuropathy may occur in a dose- and schedule-dependent fashion.

### 1.3.3 Docetaxel

***Characteristics and efficacy*** The impetus for the development of docetaxel resulted from the overharvesting of the Pacific yew. Docetaxel is a semisynthetic derivative of 10-deacetyl baccatin III, which is extracted from the needles of the European yew tree (*Taxus baccata*) (Verweij et al. 1994). Docetaxel uses Tween 80 and alcohol as a diluent. Its general mechanism of action mirrors that of paclitaxel: it promotes microtubule assembly and inhibits depolymerization, preventing proper mitotic spindle function and cell division. MAPs such as MAP2 and Tau may serve as cofactors that determine docetaxel sensitivity (Fromes et al. 1996).

***Dosing*** Patients receiving docetaxel may be premedicated with corticosteroids (e.g., dexamethasone 8 mg BID) beginning 1 day prior to administration to prevent fluid retention and hypersensitivity reactions. The optimal dosing of docetaxel in ovarian cancers is not strictly defined. The following administrations have been tolerated: 50 mg/m$^2$/day on days 1 and 8 every 3 weeks; 70–90 mg/m$^2$ by 24-h continuous IV infusion every 3 weeks; 80–100 mg/m$^2$ by 6-h infusion every 3 weeks; 100 mg/m2 by 1-h infusion every 3 weeks; 14 mg/m$^2$/day for 5 days by 1-h infusion every 21 days (Markman et al. 2001; Kushner et al. 2007; Chi et al. 2017). The dosing in the SCOTROC was 75 mg/m$^2$ IV over 1 h every 3 weeks for 6 cycles (Vasey et al. 2004). Patients with moderate hepatic impairment should receive 75 mg/m$^2$ over 1 h, while no safe dose exists for patients with greater than moderate impairment.

***Efficacy*** The phase III randomized SCOTROC (Scottish Randomised Trial in Ovarian Cancer) studied docetaxel-carboplatin versus paclitaxel-carboplatin as first-line chemotherapy for epithelial ovarian carcinoma and demonstrated similar PFS (15 vs. 14.8 months), OS (64.2% vs 68.9%), and tumor response rates (58.7% vs 59.5%). This suggests a role for docetaxel in primary treatment of epithelial ovarian cancer. The notable difference between the two regimens was their toxicity profiles, as discussed below.

***Toxicity*** Neutropenia is the dose-limiting toxicity of docetaxel, which generally resolves within 7–8 days. Other grade 3–4 toxicities include stomatitis (0–5%), dermatitis (6–20%), acute hypersensitivity reaction (7–12%), fluid retention (8–12%), and diarrhea (6–20%). Other potential side effects include alopecia, anemia, neurosensory effects, and weakness. As previously discussed, premedication with steroids decreases fluid retention. Compared to paclitaxel, docetaxel carries a higher risk for neutropenia, hypersensitivity reactions, and nausea and vomiting (Vasey et al. 2004). Paclitaxel carries higher risk for neuropathy, weakness, and myalgias. As such, the decision to administer one versus the other may be individualized based on their toxicity profiles (Tymon-Rosario et al. 2021).

Patients with impaired liver function have a significant reduction in docetaxel clearance. Patients with moderate impairment in liver function (transaminases 1.5 times the upper limit of normal and alkaline phosphatase 2.5 times the upper limit of normal) have a 27% reduction in docetaxel clearance. Patients with moderately impaired liver function have a greater incidence of febrile neutropenia compared to patients with normal liver function (40% vs 16%). Other grade 3 to 4 toxicities are increased in patients with impaired liver function, including thrombocytopenia, stomatitis, nausea, and vomiting.

### 1.3.4 Ixabepilone

***Characteristics and efficacy*** Epothilones (epoxide-thiazole-ketones) (Parajuli et al. 2014) are derived from the soil bacterium *Soarngium serosum*. Ixabepilone (Ixempra®, R-Pharm, LLC, Princeton, NJ) is a semi-synthetic analog of the naturally occurring epothilone B. Similar to paclitaxel, ixabepilone is poorly water soluble and therefore diluted in Cremophor® EL or polyoxyethylated castor oil and contraindicated in the setting of hypersensitivity to either of these agents. In GOG-126M among patients with platinum- and taxane-resistant ovarian cancer, weekly monotherapy produced an ORR of 14.3%, with 40.8% achieving stable disease. Median time to progression was 4.4 months with a median OS of 14.8 months (De Geest et al. 2010). In patients with platinum-resistant or -refractory disease who received the combination of weekly ixabepilone/biweekly bevacizumab compared to ixabepilone alone, the ORR was 38% versus 8% ($p = 0.003$) (Roque et al. 2024). A durable response at 6 months was achieved in 37% versus 3%, respectively ($p < 0.001$). The addition of bevacizumab significantly improved PFS (median: 5.5 vs 2.2 months, HR = 0.33, 95% CI 0.19–0.55, $p < 0.001$) and OS (median: 10.0 vs 6.0 months, HR = 0.52, 95% CI 0.31–0.87, $P = 0.006$) (Roque et al. 2022). A phase II study of ixabepilone for cisplatin-refractory germ cell tumors (that included only one patient with ovarian origin) revealed responses only in patients with no prior taxane exposure (Feldman et al. 2007).

***Dosing*** The FDA-approved (Food and Drug Administration 2009) indication for ixabepilone is as follows: (1) in combination with capecitabine for the treatment of metastatic or locally advanced breast cancer in patients whose tumors are resistant to treatment with an anthracycline and a taxane, or whose cancer is taxane resistant and for whom further anthracycline therapy is contraindicated and (2) as monotherapy for the treatment of metastatic or locally advanced breast cancer in patients whose tumors are resistant or refractory to anthracyclines, taxanes, and capecitabine. Ixabepilone was first examined in phase II testing for ovarian cancer in 2010 at 20 mg/m$^2$ on days 1, 8, 15 of a 28-day cycle as monotherapy in GOG-126M (De Geest et al. 2010) and then in combination with bevacizumab at 10 mg/kg on days 1 and 15 (Roque et al. 2022).

***Toxicity*** As a monotherapy, adverse effects include peripheral grade 2 (28.5%) and grade 3 (6.1%) neuropathy, grades 3 to 4 neutropenia (20.4%), grade 3 fatigue (14.3%), grade 3 nausea/emesis (22%), grade 3 diarrhea (10%), and grade 3 mucositis (4%). In 2023, the NCCN (National Comprehensive Cancer Network 2023) approved weekly ixabepilone 20 mg/m$^2$ on days 1, 8, 15 in combination with biweekly bevacizumab 10 mg/kg on days 1, 15 (category 2B) (Roque et al. 2021). Hypertension was more common in participants who received bevacizumab (36%) compared to ixabepilone alone (8%) ($p = 0.005$); most cases were grade 1–2. Participants in this group were also more likely to report sensory neuropathy (51% versus 19%, $p = 0.004$) and musculoskeletal (56% versus $n = 9$, $p = 0.006$). The rate of bowel perforation was 2.5%, which is consistent with existing literature for bevacizumab. A dose reduction by 20% was necessary in 59% of patients, though the study population was heavily pre-treated, with approximately half in either arm in receipt of a median of >3 prior lines.

## 1.4 Microtubule-Destabilizing Agents

### 1.4.1 Vincristine

***Characteristics*** Vincristine is a natural product vinca alkaloid isolated from the Madagascar periwinkle (*Catharanthus roseus*) (Moudi et al. 2013). It functions as a destabilizer of microtubule polymerization by interacting with tubulin and preventing addition of subunits, thwarting construction of a functional mitotic spindle that results in mitotic arrest. As part of the VAC (vincristine, actinomycin-D, cyclophosphamide) regimen, vincristine is currently indicated for salvage therapy in recurrent germ cell and sex cord stromal tumor (National Comprehensive Cancer Network 2023).

***Dosing*** As part of the VAC regimen, vincristine is administered at 1.5 mg/m$^2$ (to a maximum dose of 2 mg) once weekly for 8 to 12 weeks in combination with dactinomycin and cyclophosphamide (Williams et al. 1989; Slayton et al. 1985).

***Efficacy*** Importantly, no randomized trials have been conducted in this disease population. An early GOG study demonstrated decreased likelihood of disease progression in patients treated with VAC who had complete surgical resection of their disease prior to receiving the regimen (28% vs 68%) (Slayton et al. 1985).

***Toxicity*** Neurotoxicity is the most common side effect of vincristine (Madsen et al. 2019). Indeed, vincristine is the most neurotoxic vinca alkaloid and virtually all patients experience some degree of neuropathy (Legha 1986). Additionally, toxicities include gastrointestinal upset, alopecia, myelosuppression, and hypersensitivity reactions.

#### 1.4.1.1 Vinblastine

***Characteristics*** Like its analog vincristine, vinblastine is a natural product of the Madagascar periwinkle that acts as a microtubule destabilizer. It is a possible treatment for recurrent malignant ovarian germ cell tumors as part of VeIP (in combination with ifosfamide and cisplatin) (National Comprehensive Cancer Network 2023).

***Dosing*** Vinblastine 0.11 mg/kg/day IV on days 1 and 2 (in combination with cisplatin 20 mg/m$^2$ on days 1–5, ifosfamide 1.2 g/m$^2$ on days 1–5, and mesna) every 21 days for a total of 4 cycles.

***Efficacy*** A phase II trial examining VeIP in recurrent germ cell tumors (in a primarily male population with only one female patient) demonstrated an ORR of 49.6% with 23.7% remaining disease-free at 6 years follow-up (Loehrer et al. 1998).

***Toxicity*** Vinblastine is associated with peripheral neuropathy, but less severe than that of vincristine. The major dose-limiting toxicity is hematologic, including leukopenia, anemia, and thrombocytopenia. Other toxicities include alopecia, constipation, and hypertension.

### 1.4.2 Vinorelbine

***Characteristics*** Vinorelbine is a semisynthetic third-generation vinca alkaloid. Its chemical structure differs from that of vincristine and vinblastine by the addition of a catharanthine ring to reduce neurotoxic side effects. Its mechanism of action is similar to that of other vinca alkaloids, namely destabilization of microtubules. It is recommended as single-agent therapy for patients with recurrent platinum-resistant epithelial ovarian cancer (National Comprehensive Cancer Network 2023).

***Dosing*** Dosing varies between studies, but the typical dosing is 30 mg/m$^2$ weekly, though this is decreased to 25 mg/m$^2$ weekly if it is combined with cisplatin (100 mg/m$^2$ every 4 weeks). Of note, vinorelbine is a vesicant and must be handled with caution during administration. Following administration, the vein should be flushed with at least 75 mL IV solution.

***Efficacy*** Vinorelbine has been studied as salvage therapy in women with recurrent ovarian cancer with mixed results. ORR range from 3% to 29%. A phase II study in patients with recurrent ovarian cancer demonstrated an ORR of 15.2%, all of which were in platinum-resistant patients; however, only 30% of these patients had received paclitaxel (Bajetta et al. 1996). A subsequent phase II study evaluating weekly vinorelbine as salvage therapy in women with recurrent or resistant epithelial ovarian cancer following platinum and paclitaxel therapy showed a poor response rate of 3% and substantial degree of symptom-related distress as measured by the Symptom Distress Scale (Rothenberg et al. 2004).

***Toxicity*** Myelosuppression, most specifically granulocytopenia, is the primary dose-limiting toxicity of vinorelbine, present in 36% of patients. This effect is non-cumulative, and the granulocyte nadir occurred on day 14 of treatment. Compared to vincristine, vinorelbine is associated with relatively mild peripheral neuropathy. Other toxicities include constipation, local extravasation (leading to irritation, tissue necrosis, and/or thrombophlebitis), and pulmonary toxicity.

## 1.5 Mechanisms of Resistance

While nearly 80% of newly diagnosed patients respond to initial treatment (Ozols et al. 2003), drug resistance eventually ensues, to which patients eventually succumb. Common mechanisms include increased efflux of drug, alterations in drug binding affinity, compensatory pathway alterations (e.g., leading to defective apoptosis, overactivation of growth pathways such as PI3K/AKT, enhanced detoxification by glutathione), changes in other regulators of tubulin (Ortiz et al. 2022).

***Drug Efflux*** Resistance to both taxanes and vinca alkaloids may be mediated by efflux pumps. The natural function of efflux pumps is to protect the intracellular environment from potentially harmful chemicals by transporting them back to the extracellular space. The most common family of efflux pumps are ABC (ATP-binding cassette) transporters, a superfamily of ATP-dependent translocases. The best categorized and most common mechanism of these is P-glycoprotein (P-gp, also known as MDR1), encoded by the *MDR1* gene. Overexpression of MDR1 leads to both taxane and vinca alkaloid resistance (Yusuf et al. 2003; Ambudkar et al. 1999). The mechanism of resistance by P-gp is simple: reduction of intracellular drug concentrations. Intracellular drug binding induces ATP hydrolysis and produces a conformational change that translocates the drug to the extracellular space (Ramachandra et al. 1998). P-gp overexpression often also functions as a significant negative predictor of ovarian cancer survival and paclitaxel response (Penson et al. 2004). Several other ABC proteins, including MRP1 (ABCC1) and BRCP (ABCG2), function similarly to P-gp to produce resistance to other chemotherapeutics (Fojo and Menefee 2007).

***Alterations in binding site affinity*** At least eight α-tubulin and seven β-tubulin genes have been identified in humans (Table 1.1) that diverge at the C-terminus (Janke and Magiera 2020; Verdier-Pinard et al. 2009). Paclitaxel preferentially binds to class I β-tubulin (Magnani et al. 2006). Chemoresistant ovarian cancers may upregulate class III β-tubulin (Roque et al. 2013, 2014; Ferlini et al. 2007), a tubulin isotype that differs from class I β-tubulin at the paclitaxel binding site. Epothilones are touted as agents that can overcome taxane resistance by retaining their binding affinity despite upregulation of class III β-tubulin. Overexpression of class III β-tubulin may also reduce microtubule assembly (Hari et al. 2003).

***Apoptotic signaling deficiency*** Evasion of apoptosis is a hallmark of cancer (Hanahan and Weinberg 2011). Bcl-2 is a critical upstream negative regulator of apoptosis that can be phosphorylated to induce apoptosis. Paclitaxel-induced mitotic arrest initiates Bcl-2 phosphorylation, contributing to cell death (Bhalla 2003). Bcl-2 overexpression is associated with ovarian cancer resistance to paclitaxel (Srivastava et al. 1999). Additionally, Bcl-2 family proteins (including the pro-survival proteins Mcl-1 and Bcl-xL, as well as the pro-apoptotic proteins BH3-only, BAX, BAK, and BOK) have been shown to be differentially regulated in ovarian cancer (Yuan et al. 2020). Consequently, differential expression of Bcl-2 family proteins (especially overexpression of Bcl-2 itself) provides an important mechanism of resistance for ovarian cancer cells.

***Dysregulation by MAPs*** Numerous microtubule-associated proteins (discussed in detail in the following chapter) are differentially regulated in ovarian cancer cells. Tau overexpression is associated with resistance to paclitaxel,

**Table 1.1** Alpha and beta tubulin isotypes (Verdier-Pinard et al. 2009; Ludueña 2013; Ferreira et al. 2018)

| Tubulin Isotype | Reference Sequence (National Center for Biotechnology Information database, www.ncbi.nlm.nih.gov) | C-Terminal Sequence | Tissue distribution |
|---|---|---|---|
| α1A | **NP_006000** | MAALEKDYEEVGVDSVEGEGEEEGEEY | Ubiquitous |
| α 1B | **AAC31959** | MAALEKDYEEVGVDSVEGEGEEEGEEY | Ubiquitous |
| α 1C | **Q9BQE3** | MAALEKDYEEVGADSADGEDEGEEY | Ubiquitous |
| α 3C | P0DPH7.1 | LAALEKDYEEVGVDSVEAEAEEGEEY | Enriched in testis |
| α 3E | **NP_997195** | LAALEKDCEEVGVDSVEAEAEEGEEY | Enriched in testis |
| α 4A | **NP_005991** | MAALEKDYEEVG I D SYEDEDEGEE | Enriched in testis |
| α 8 | **Q9NY65** | LAALEKDYEEVGTDSFEEENEGEEF | Ubiquitous |
| α -like 3 | **NP_079079** | LAALERDYEEVAQSF | Ubiquitous |
| Bα1 | **CAA25855** | MAALEKDYEEVGVHSVEGEGEEEGEEY | Ubiquitous |
| βI | **AAD33873** | YQDATAEEEEDFGEEAEEEA | Ubiquitous |
| βII | **NP 001060/AAH01352** | YQDATADEQGEFEEEEGEDEA | Neurons, glia, retina, small intestine |
| βIII | **AAH00748** | YQDATAEEEGEMYEDDEEESEAQGPK | Neurons, testis (Sertoli cells and sperm cells), vestibular organs, nasal epithelia, colon |
| βIVa | **NP_006078** | YQDATAEQGEFEEEAEEEVAYQDATAEEGEFEEEAEEEVA | Ciliated cells: retinal rods, sperm flagella, and the cilia of trachea, oviduct, vestibular hair cells, olfactory neurons, brain, efferent duct of the testis, heart, skeletal muscle; IVA brain, IVB testis, heart, skeletal muscle |
| βIVb | **CAA26203** | YQDATAEEEGEFEEEAEEEVA | |
| βV | **NP_115914** | YQDATANDGEEAFEDEEEE I DG | Skin, skeletal, smooth and cardiac muscles, endothelial cells, bile duct, lymph nodes, spleen, breast, islets of Langerhans, testis, tonsil, placenta, prostate, fetal liver; mostly absent from the brain, thymus, bone marrow, leukocytes |
| βVI | **NP_110400** | FQDAKAVLEEDEEVTEEAEMEPEDKGH | Hematopoietic cells, chicken erythrocytes, mammalian spleen, platelets and mega- karyocytes |

possibly by outcompeting paclitaxel for microtubule binding sites (Smoter et al. 2011). Additionally, overexpression of survivin or stathmin is associated with resistance to paclitaxel and poor prognosis (Cohen et al. 2003; Su et al. 2009). By contrast, CLIP170 expression is associated with paclitaxel sensitivity (Sun et al. 2012). MAP4 increases microtubule polymerization, sensitivity to paclitaxel, and paclitaxel binding while decreasing sensitivity to vinca alkaloids (Zhang et al. 1998).

## References


Akhmanova A, Steinmetz MO. Control of microtubule organization and dynamics: two ends in the limelight. Nat Rev Mol Cell Biol 2015;16:711–26. https://doi.org/10.1038/nrm4084.

Alberts B, Heald R, Johnson A, Morgan D, Raff M, Roberts K, et al. Molecular Biology of the Cell. 7th ed. Garland; 2017.

Ambudkar SV, Dey S, Hrycyna CA, Ramachandra M, Pastan I, Gottesman MM. Biochemical, cellular, and pharmacological aspects of the multidrug transporter. Annu Rev Pharmacol Toxicol 1999;39:361–98. https://doi.org/10.1146/annurev.pharmtox.39.1.361.

Armstrong DK, Bundy B, Wenzel L, Huang HQ, Baergen R, Lele S, et al. Intraperitoneal cisplatin and paclitaxel in ovarian cancer. N Engl J Med 2006;354:34–43. https://doi.org/10.1056/NEJMoa052985.

Bajetta E, Di Leo A, Biganzoli L, Mariani L, Cappuzzo F, Di Bartolomeo M, et al. Phase II study of vinorelbine in patients with pretreated advanced ovarian cancer: activity in platinum-resistant disease. J Clin Oncol 1996;14:2546–51. https://doi.org/10.1200/JCO.1996.14.9.2546.

Barroso-Sousa R, Vaz-Luis I, Di Meglio A, Hu J, Li T, Rees R, et al. Prospective Study Testing a Simplified Paclitaxel Premedication Regimen in Patients with Early Breast Cancer. The Oncologist 2021;26:927–33. https://doi.org/10.1002/onco.13960.

Bhalla KN. Microtubule-targeted anticancer agents and apoptosis. Oncogene 2003;22:9075–86. https://doi.org/10.1038/sj.onc.1207233.

Brouhard GJ, Rice LM. Microtubule dynamics: an interplay of biochemistry and mechanics. Nat Rev Mol Cell Biol 2018;19:451–63. https://doi.org/10.1038/s41580-018-0009-y.

Brown J, Shvartsman HS, Deavers MT, Burke TW, Munsell MF, Gershenson DM. The Activity of Taxanes in the Treatment of Sex Cord-Stromal Ovarian Tumors. J Clin Oncol 2004;22:3517–23. https://doi.org/10.1200/JCO.2004.12.074.

Brown J, Shvartsman HS, Deavers MT, Ramondetta LM, Burke TW, Munsell MF, et al. The activity of taxanes compared with bleomycin, etoposide, and cisplatin in the treatment of sex cord-stromal ovarian tumors. Gynecol Oncol 2005;97:489–96. https://doi.org/10.1016/j.ygyno.2005.01.011.

Buster DW, Zhang D, Sharp DJ. Poleward tubulin flux in spindles: regulation and function in mitotic cells. Mol Biol Cell 2007;18:3094–104. https://doi.org/10.1091/mbc.e06-11-0994.

Carter AP, Cho C, Jin L, Vale RD. Crystal structure of the dynein motor domain. Science 2011;331:1159–65. https://doi.org/10.1126/science.1202393.

Carvalho-Santos Z, Azimzadeh J, Pereira-Leal JB, Bettencourt-Dias M. Evolution: Tracing the origins of centrioles, cilia, and flagella. J Cell Biol 2011;194:165–75. https://doi.org/10.1083/jcb.201011152.

Cassimeris L, Spittle C. Regulation of microtubule-associated proteins. Int Rev Cytol 2001;210:163–226. https://doi.org/10.1016/s0074-7696(01)10006-9.

Chan JK, Brady MF, Penson RT, Huang H, Birrer MJ, Walker JL, et al. Weekly vs. Every-3-Week Paclitaxel and Carboplatin for Ovarian Cancer. N Engl J Med 2016;374:738–48. https://doi.org/10.1056/NEJMoa1505067.

Chandrasekaran G, Tátrai P, Gergely F. Hitting the brakes: targeting microtubule motors in cancer. Br J Cancer 2015;113:693–8. https://doi.org/10.1038/bjc.2015.264.

Chen N, Li Y, Ye Y, Palmisano M, Chopra R, Zhou S. Pharmacokinetics and pharmacodynamics of nab-paclitaxel in patients with solid tumors: disposition kinetics and pharmacology distinct from solvent-based paclitaxel. J Clin Pharmacol 2014;54:1097–107. https://doi.org/10.1002/jcph.304.

Chen W, Zhang D. Kinetochore fibre dynamics outside the context of the spindle during anaphase. Nat Cell Biol 2004;6:227–31. https://doi.org/10.1038/ncb1104.

Chi D, Berchuck A, Dizon D, Yashar C. Principles and Practice of Gynecologic Oncology. 7th ed. Lippincott Williams & Wilkins; 2017.

Clamp AR, James EC, McNeish IA, Dean A, Kim J-W, O'Donnell DM, et al. Weekly dose-dense chemotherapy in first-line epithelial ovarian, fallopian tube, or primary peritoneal carcinoma treatment (ICON8): primary progression free survival analysis results from a GCIG phase 3 randomised controlled trial. Lancet Lond Engl 2019;394:2084–95. https://doi.org/10.1016/S0140-6736(19)32259-7.

Cohen C, Lohmann CM, Cotsonis G, Lawson D, Santoianni R. Survivin expression in ovarian carcinoma: correlation with apoptotic markers and prognosis. Mod Pathol Off J U S Can Acad Pathol Inc 2003;16:574–83. https://doi.org/10.1097/01.MP.0000073868.31297.B0.

Coleman RL, Brady WE, McMeekin DS, Rose PG, Soper JT, Lentz SS, et al. A phase II evaluation of nanoparticle, albumin-bound (nab) paclitaxel in the treatment of recurrent or persistent platinum-resistant ovarian, fallopian tube, or primary peritoneal cancer: a Gynecologic Oncology Group study. Gynecol

Oncol 2011;122:111–5. https://doi.org/10.1016/j.ygyno.2011.03.036.

Conde C, Cáceres A. Microtubule assembly, organization and dynamics in axons and dendrites. Nat Rev Neurosci 2009;10:319–32. https://doi.org/10.1038/nrn2631.

Courtois A, Schuh M, Ellenberg J, Hiiragi T. The transition from meiotic to mitotic spindle assembly is gradual during early mammalian development. J Cell Biol 2012;198:357–70. https://doi.org/10.1083/jcb.201202135.

De Geest K, Blessing JA, Morris RT, Yamada SD, Monk BJ, Zweizig SL, et al. Phase II clinical trial of ixabepilone in patients with recurrent or persistent platinum- and taxane-resistant ovarian or primary peritoneal cancer: a gynecologic oncology group study. J Clin Oncol Off J Am Soc Clin Oncol 2010;28:149–53. https://doi.org/10.1200/JCO.2009.24.1455.

Desai A, Mitchison TJ. Microtubule polymerization dynamics. Annu Rev Cell Dev Biol 1997;13:83–117. https://doi.org/10.1146/annurev.cellbio.13.1.83.

Domnitz SB, Wagenbach M, Decarreau J, Wordeman L. MCAK activity at microtubule tips regulates spindle microtubule length to promote robust kinetochore attachment. J Cell Biol 2012;197:231–7. https://doi.org/10.1083/jcb.201108147.

Dráber P, Dráberová E. Dysregulation of Microtubule Nucleating Proteins in Cancer Cells. Cancers 2021;13:5638. https://doi.org/10.3390/cancers13225638.

Eisenhauer EA, ten Bokkel Huinink WW, Swenerton KD, Gianni L, Myles J, van der Burg ME, et al. European-Canadian randomized trial of paclitaxel in relapsed ovarian cancer: high-dose versus low-dose and long versus short infusion. J Clin Oncol Off J Am Soc Clin Oncol 1994;12:2654–66. https://doi.org/10.1200/JCO.1994.12.12.2654.

Fagotti A, Ferrandina G, Vizzielli G, Fanfani F, Gallotta V, Chiantera V, et al. Phase III randomised clinical trial comparing primary surgery versus neoadjuvant chemotherapy in advanced epithelial ovarian cancer with high tumour load (SCORPION trial): Final analysis of peri-operative outcome. Eur J Cancer Oxf Engl 1990 2016;59:22–33. https://doi.org/10.1016/j.ejca.2016.01.017.

Feldman DR, Kondagunta GV, Ginsberg MS, Ishill N, Patil S, Cestaro J, et al. Phase II trial of ixabepilone in patients with cisplatin-refractory germ cell tumors. Invest New Drugs 2007;25:487–90. https://doi.org/10.1007/s10637-007-9059-2.

Ferlini C, Raspaglio G, Cicchillitti L, Mozzetti S, Prislei S, Bartollino S, et al. Looking at drug resistance mechanisms for microtubule interacting drugs: does TUBB3 work? Curr Cancer Drug Targets 2007;7:704–12.

Ferreira LT, Figueiredo AC, Orr B, Lopes D, Maiato H. Dissecting the role of the tubulin code in mitosis. Methods Cell Biol 2018;144:33–74. https://doi.org/10.1016/bs.mcb.2018.03.040.

Field JJ, Díaz JF, Miller JH. The binding sites of microtubule-stabilizing agents. Chem Biol 2013;20:301–15. https://doi.org/10.1016/j.chembiol.2013.01.014.

Fojo T, Menefee M. Mechanisms of multidrug resistance: the potential role of microtubule-stabilizing agents. Ann Oncol Off J Eur Soc Med Oncol 2007;18 Suppl 5:v3–8. https://doi.org/10.1093/annonc/mdm172.

Food and Drug Administration. TAXOL (paclitaxel) INJECTION (reference ID: 2939751). 2011.

Food and Drug Administration. Ixabepilone: highlights of prescribing information. 2009.

de Forges H, Bouissou A, Perez F. Interplay between microtubule dynamics and intracellular organization. Int J Biochem Cell Biol 2012;44:266–74. https://doi.org/10.1016/j.biocel.2011.11.009.

Fromes Y, Gounon P, Veitia R, Bissery MC, Fellous A. Influence of microtubule-associated proteins on the differential effects of paclitaxel and docetaxel. J Protein Chem 1996;15:377–88. https://doi.org/10.1007/BF01886864.

Fu M, Holzbaur ELF. Integrated regulation of motor-driven organelle transport by scaffolding proteins. Trends Cell Biol 2014;24:564–74. https://doi.org/10.1016/j.tcb.2014.05.002.

Gadde S, Heald R. Mechanisms and molecules of the mitotic spindle. Curr Biol CB 2004;14:R797–805. https://doi.org/10.1016/j.cub.2004.09.021.

Ganem NJ, Compton DA. Functional roles of poleward microtubule flux during mitosis. Cell Cycle Georget Tex 2006;5:481–5. https://doi.org/10.4161/cc.5.5.2519.

Gershenson DM, Okamoto A, Ray-Coquard I. Management of Rare Ovarian Cancer Histologies. J Clin Oncol Off J Am Soc Clin Oncol 2019;37:2406–15. https://doi.org/10.1200/JCO.18.02419.

Gibbons IR, Lee-Eiford A, Mocz G, Phillipson CA, Tang WJ, Gibbons BH. Photosensitized cleavage of dynein heavy chains. Cleavage at the "V1 site" by irradiation at 365 nm in the presence of ATP and vanadate. J Biol Chem 1987;262:2780–6.

Goodson HV, Jonasson EM. Microtubules and Microtubule-Associated Proteins. Cold Spring Harb Perspect Biol 2018;10:a022608. https://doi.org/10.1101/cshperspect.a022608.

Goulet A, Moores C. New insights into the mechanism of force generation by kinesin-5 molecular motors. Int Rev Cell Mol Biol 2013;304:419–66. https://doi.org/10.1016/B978-0-12-407696-9.00008-7.

Gynecologic Oncology Group, Markman M, Blessing J, Rubin SC, Connor J, Hanjani P, et al. Phase II trial of weekly paclitaxel (80 mg/m2) in platinum and paclitaxel-resistant ovarian and primary peritoneal cancers: a Gynecologic Oncology Group study. Gynecol Oncol 2006;101:436–40. https://doi.org/10.1016/j.ygyno.2005.10.036.

Hájek R, Vorlicek J, Slavik M. Paclitaxel (Taxol): a review of its antitumor activity in clinical studies Minireview. Neoplasma 1996;43:141–54.

Hanahan D, Weinberg RA. Hallmarks of cancer: the next generation. Cell 2011;144:646–74. https://doi.org/10.1016/j.cell.2011.02.013.

Hari M, Yang H, Zeng C, Canizales M, Cabral F. Expression of class III beta-tubulin reduces microtubule assembly and confers resistance to paclitaxel. Cell Motil Cytoskeleton 2003;56:45–56. https://doi.org/10.1002/cm.10132.

Hinton S, Catalano P, Einhorn LH, Loehrer PJ, Kuzel T, Vaughn D, et al. Phase II study of paclitaxel plus gemcitabine in refractory germ cell tumors (E9897): a trial of the Eastern Cooperative Oncology Group. J Clin Oncol Off J Am Soc Clin Oncol 2002;20:1859–63. https://doi.org/10.1200/JCO.2002.07.158.

Holzbaur EL, Johnson KA. Microtubules accelerate ADP release by dynein. Biochemistry 1989;28:7010–6. https://doi.org/10.1021/bi00443a034.

Janke C, Magiera MM. The tubulin code and its role in controlling microtubule properties and functions. Nat Rev Mol Cell Biol 2020;21:307–26. https://doi.org/10.1038/s41580-020-0214-3.

Jiang C, You Q. Kinesin spindle protein inhibitors in cancer: a patent review (2008 - present). Expert Opin Ther Pat 2013;23:1547–60. https://doi.org/10.1517/13543776.2013.833606.

Johnson KB, Sharma A, Henry NL, Wei M, Bie B, Hershberger CE, et al. Genetic variations that influence paclitaxel pharmacokinetics and intracellular effects that may contribute to chemotherapy-induced neuropathy: A narrative review. Front Pain Res 2023;4:1139883. https://doi.org/10.3389/fpain.2023.1139883.

Kaan HYK, Hackney DD, Kozielski F. The structure of the kinesin-1 motor-tail complex reveals the mechanism of autoinhibition. Science 2011;333:883–5. https://doi.org/10.1126/science.1204824.

Katsumata N, Yasuda M, Takahashi F, Isonishi S, Jobo T, Aoki D, et al. Dose-dense paclitaxel once a week in combination with carboplatin every 3 weeks for advanced ovarian cancer: a phase 3, open-label, randomised controlled trial. Lancet Lond Engl 2009;374:1331–8. https://doi.org/10.1016/S0140-6736(09)61157-0.

Kehoe S, Hook J, Nankivell M, Jayson GC, Kitchener H, Lopes T, et al. Primary chemotherapy versus primary surgery for newly diagnosed advanced ovarian cancer (CHORUS): an open-label, randomised, controlled, non-inferiority trial. Lancet Lond Engl 2015;386:249–57. https://doi.org/10.1016/S0140-6736(14)62223-6.

Kollman JM, Merdes A, Mourey L, Agard DA. Microtubule nucleation by γ-tubulin complexes. Nat Rev Mol Cell Biol 2011;12:709–21. https://doi.org/10.1038/nrm3209.

Kon T, Nishiura M, Ohkura R, Toyoshima YY, Sutoh K. Distinct functions of nucleotide-binding/hydrolysis sites in the four AAA modules of cytoplasmic dynein. Biochemistry 2004;43:11266–74. https://doi.org/10.1021/bi048985a.

Kondagunta GV, Bacik J, Donadio A, Bajorin D, Marion S, Sheinfeld J, et al. Combination of paclitaxel, ifosfamide, and cisplatin is an effective second-line therapy for patients with relapsed testicular germ cell tumors. J Clin Oncol Off J Am Soc Clin Oncol 2005;23:6549–55. https://doi.org/10.1200/JCO.2005.19.638.

Kuroi K, Shimozuma K. Neurotoxicity of taxanes: symptoms and quality of life assessment. Breast Cancer Tokyo Jpn 2004;11:92–9. https://doi.org/10.1007/BF02968010.

Kushner DM, Connor JP, Sanchez F, Volk M, Schink JC, Bailey HH, et al. Weekly docetaxel and carboplatin for recurrent ovarian and peritoneal cancer: a phase II trial. Gynecol Oncol 2007;105:358–64. https://doi.org/10.1016/j.ygyno.2006.12.018.

Lansinger OM, Biedermann S, He Z, Colevas AD. Do Steroids Matter? A Retrospective Review of Premedication for Taxane Chemotherapy and Hypersensitivity Reactions. J Clin Oncol Off J Am Soc Clin Oncol 2021;39:3583–90. https://doi.org/10.1200/JCO.21.01200.

Legha SS. Vincristine neurotoxicity. Pathophysiology and management. Med Toxicol 1986;1:421–7. https://doi.org/10.1007/BF03259853.

Loehrer PJ, Gonin R, Nichols CR, Weathers T, Einhorn LH. Vinblastine plus ifosfamide plus cisplatin as initial salvage therapy in recurrent germ cell tumor. J Clin Oncol Off J Am Soc Clin Oncol 1998;16:2500–4. https://doi.org/10.1200/JCO.1998.16.7.2500.

Ludueña RF. A hypothesis on the origin and evolution of tubulin. Int Rev Cell Mol Biol 2013;302:41–185. https://doi.org/10.1016/B978-0-12-407699-0.00002-9.

Madsen ML, Due H, Ejskjær N, Jensen P, Madsen J, Dybkær K. Aspects of vincristine-induced neuropathy in hematologic malignancies: a systematic review. Cancer Chemother Pharmacol 2019;84:471–85. https://doi.org/10.1007/s00280-019-03884-5.

Magnani M, Ortuso F, Soro S, Alcaro S, Tramontano A, Botta M. The betaI/betaIII-tubulin isoforms and their complexes with antimitotic agents. Docking and molecular dynamics studies. FEBS J 2006;273:3301–10. https://doi.org/10.1111/j.1742-4658.2006.05340.x.

Margolis RL, Wilson L. Microtubule treadmills—possible molecular machinery. Nature 1981;293:705–11. https://doi.org/10.1038/293705a0.

Markman M, Kennedy A, Webster K, Kulp B, Peterson G, Belinson J. Paclitaxel-associated hypersensitivity reactions: experience of the gynecologic oncology program of the Cleveland Clinic Cancer Center. J Clin Oncol Off J Am Soc Clin Oncol 2000;18:102–5. https://doi.org/10.1200/JCO.2000.18.1.102.

Markman M, Kennedy A, Webster K, Peterson G, Kulp B, Belinson J. Combination chemotherapy with carboplatin and docetaxel in the treatment of cancers of the ovary and fallopian tube and primary carcinoma of the peritoneum. J Clin Oncol Off J Am Soc Clin Oncol 2001;19:1901–5. https://doi.org/10.1200/JCO.2001.19.7.1901.

Marupudi NI, Han JE, Li KW, Renard VM, Tyler BM, Brem H. Paclitaxel: a review of adverse toxicities and novel delivery strategies. Expert

Opin Drug Saf 2007;6:609–21. https://doi.org/10.1517/14740338.6.5.609.

Marx A, Hoenger A, Mandelkow E. Structures of kinesin motor proteins. Cell Motil Cytoskeleton 2009;66:958–66. https://doi.org/10.1002/cm.20392.

Maurer K, Michener C, Mahdi H, Rose PG. Universal tolerance of nab-paclitaxel for gynecologic malignancies in patients with prior taxane hypersensitivity reactions. J Gynecol Oncol 2017;28:e38. https://doi.org/10.3802/jgo.2017.28.e38.

McGuire WP, Hoskins WJ, Brady MF, Kucera PR, Partridge EE, Look KY, et al. Cyclophosphamide and cisplatin compared with paclitaxel and cisplatin in patients with stage III and stage IV ovarian cancer. N Engl J Med 1996;334:1–6. https://doi.org/10.1056/NEJM199601043340101.

McIntosh JR. Mitosis. Cold Spring Harb Perspect Biol 2016;8:a023218. https://doi.org/10.1101/cshperspect.a023218.

Mitchison TJ. Polewards microtubule flux in the mitotic spindle: evidence from photoactivation of fluorescence. J Cell Biol 1989;109:637–52. https://doi.org/10.1083/jcb.109.2.637.

Moudi M, Go R, Yien CYS, Nazre M. Vinca alkaloids. Int J Prev Med 2013;4:1231–5.

Muggia FM, Braly PS, Brady MF, Sutton G, Niemann TH, Lentz SL, et al. Phase III randomized study of cisplatin versus paclitaxel versus cisplatin and paclitaxel in patients with suboptimal stage III or IV ovarian cancer: a gynecologic oncology group study. J Clin Oncol Off J Am Soc Clin Oncol 2000;18:106–15. https://doi.org/10.1200/JCO.2000.18.1.106.

Murakami R, Matsumura N, Brown JB, Wang Z, Yamaguchi K, Abiko K, et al. Prediction of taxane and platinum sensitivity in ovarian cancer based on gene expression profiles. Gynecol Oncol 2016;141:49–56. https://doi.org/10.1016/j.ygyno.2016.02.027.

National Comprehensive Cancer Network. NCCN Clnical Practice Guidelines in Oncology: Ovarian, Fallopian Tube, Primary Peritoneal Cancer v1.2023 2023.

NRG Winter Meeting (virtual). 2021.

Ortiz M, Wabel E, Mitchell K, Horibata S. Mechanisms of chemotherapy resistance in ovarian cancer. Cancer Drug Resist Alhambra Calif 2022;5:304–16. https://doi.org/10.20517/cdr.2021.147.

Ozols RF, Bundy BN, Greer BE, Fowler JM, Clarke-Pearson D, Burger RA, et al. Phase III trial of carboplatin and paclitaxel compared with cisplatin and paclitaxel in patients with optimally resected stage III ovarian cancer: a Gynecologic Oncology Group study. J Clin Oncol Off J Am Soc Clin Oncol 2003;21:3194–200. https://doi.org/10.1200/JCO.2003.02.153.

Parajuli P, Pandey RP, Koirala N, Yoon YJ, Kim B-G, Sohng JK. Enzymatic synthesis of epothilone A glycosides. AMB Express 2014;4:31. https://doi.org/10.1186/s13568-014-0031-1.

Parmar MKB, Ledermann JA, Colombo N, du Bois A, Delaloye J-F, Kristensen GB, et al. Paclitaxel plus platinum-based chemotherapy versus conventional platinum-based chemotherapy in women with relapsed ovarian cancer: the ICON4/AGO-OVAR-2.2 trial. Lancet Lond Engl 2003;361:2099–106. https://doi.org/10.1016/s0140-6736(03)13718-x.

Penson RT, Oliva E, Skates SJ, Glyptis T, Fuller AF, Goodman A, et al. Expression of multidrug resistance-1 protein inversely correlates with paclitaxel response and survival in ovarian cancer patients: a study in serial samples. Gynecol Oncol 2004;93:98–106. https://doi.org/10.1016/j.ygyno.2003.11.053.

Pignata S, Scambia G, Katsaros D, Gallo C, Pujade-Lauraine E, De Placido S, et al. Carboplatin plus paclitaxel once a week versus every 3 weeks in patients with advanced ovarian cancer (MITO-7): a randomised, multicentre, open-label, phase 3 trial. Lancet Oncol 2014;15:396–405. https://doi.org/10.1016/S1470-2045(14)70049-X.

Poveda AM, Selle F, Hilpert F, Reuss A, Savarese A, Vergote I, et al. Bevacizumab Combined With Weekly Paclitaxel, Pegylated Liposomal Doxorubicin, or Topotecan in Platinum-Resistant Recurrent Ovarian Cancer: Analysis by Chemotherapy Cohort of the Randomized Phase III AURELIA Trial. J Clin Oncol Off J Am Soc Clin Oncol 2015;33:3836–8. https://doi.org/10.1200/JCO.2015.63.1408.

Pujade-Lauraine E, Hilpert F, Weber B, Reuss A, Poveda A, Kristensen G, et al. Bevacizumab combined with chemotherapy for platinum-resistant recurrent ovarian cancer: The AURELIA open-label randomized phase III trial. J Clin Oncol Off J Am Soc Clin Oncol 2014;32:1302–8. https://doi.org/10.1200/JCO.2013.51.4489.

Ramachandra M, Ambudkar SV, Chen D, Hrycyna CA, Dey S, Gottesman MM, et al. Human P-glycoprotein exhibits reduced affinity for substrates during a catalytic transition state. Biochemistry 1998;37:5010–9. https://doi.org/10.1021/bi973045u.

Robert M, Mathuranath PS. Tau and tauopathies. Neurol India 2007;55:11–6. https://doi.org/10.4103/0028-3886.30420.

Roberts AJ, Kon T, Knight PJ, Sutoh K, Burgess SA. Functions and mechanics of dynein motor proteins. Nat Rev Mol Cell Biol 2013;14:713–26. https://doi.org/10.1038/nrm3667.

Roberts AJ, Malkova B, Walker ML, Sakakibara H, Numata N, Kon T, et al. ATP-driven remodeling of the linker domain in the dynein motor. Struct Lond Engl 1993 2012;20:1670–80. https://doi.org/10.1016/j.str.2012.07.003.

Rodionov V, Nadezhdina E, Borisy G. Centrosomal control of microtubule dynamics. Proc Natl Acad Sci U S A 1999;96:115–20. https://doi.org/10.1073/pnas.96.1.115.

Rodionov VI, Borisy GG. Microtubule treadmilling in vivo. Science 1997;275:215–8. https://doi.org/10.1126/science.275.5297.215.

Roque DM, Bellone S, Buza N, Romani C, Cocco E, Bignotti E, et al. Class III β-tubulin overexpression in ovarian clear cell and serous carcinoma as a maker for poor overall survival after platinum/taxane chemotherapy and sensitivity to patupilone. Am J Obstet Gynecol 2013;209:62.e1–9. https://doi.org/10.1016/j.ajog.2013.04.017.

Roque DM, Buza N, Glasgow M, Bellone S, Bortolomai I, Gasparrini S, et al. Class III β-tubulin overexpression within the tumor microenvironment is a prognostic biomarker for poor overall survival in ovarian cancer patients treated with neoadjuvant carboplatin/paclitaxel. Clin Exp Metastasis 2014;31:101–10. https://doi.org/10.1007/s10585-013-9614-5.

Roque DM, Siegel E, Buza N, Bellone S, Silasi D, Huang G, et al. Randomized phase II trial of weekly ixabepilone with or without biweekly bevacizumab for platinum-resistant or refractory ovarian/fallopian tube/primary peritoneal cancer. Gynecol Oncol 2021;62:S3–60.

Roque DM, Siegel ER, Buza N, Bellone S, Silasi D-A, Huang GS, et al. Randomised phase II trial of weekly ixabepilone ± biweekly bevacizumab for platinum-resistant or refractory ovarian/fallopian tube/primary peritoneal cancer. Br J Cancer 2022;126:1695–703. https://doi.org/10.1038/s41416-022-01717-6.

Roque DM, Siegel ER, Buza N, Bellonne S, Huang GS, Andikyan V, CLark M, Azodi M, Schwartz PE, Rao GG, Xu F, Altwerger G, Ratner E, Santin AD. Randomized phase II trial of weekly ixabepilone ± biweekly bevacizumab for platinum-resistant orrefractory ovarian / fallopian tube / primary peritoneal cancer (NCT03093155): updated survival and subgroup analyses. Society of Gynecologic Oncology Annual Meeting Mar 16–18, 2024, San Diego, CA [poster].

Roselló S, Blasco I, García Fabregat L, Cervantes A, Jordan K, ESMO Guidelines Committee. Management of infusion reactions to systemic anticancer therapy: ESMO Clinical Practice Guidelines. Ann Oncol Off J Eur Soc Med Oncol 2017;28:iv100–18. https://doi.org/10.1093/annonc/mdx216.

Rothenberg ML, Liu PY, Wilczynski S, Nahhas WA, Winakur GL, Jiang CS, et al. Phase II trial of vinorelbine for relapsed ovarian cancer: a Southwest Oncology Group study. Gynecol Oncol 2004;95:506–12. https://doi.org/10.1016/j.ygyno.2004.09.004.

Rowinsky EK, Donehower RC. Paclitaxel (taxol). N Engl J Med 1995; 332(15):1004–14. https://doi.org/10.1056/NEJM199504133321507

Rubin SC, Randall TC, Armstrong KA, Chi DS, Hoskins WJ. Ten-year follow-up of ovarian cancer patients after second-look laparotomy with negative findings. Obstet Gynecol 1999;93:21–4. https://doi.org/10.1016/s0029-7844(98)00334-2.

Schiff PB, Fant J, Horwitz SB. Promotion of microtubule assembly in vitro by taxol. Nature 1979;277:665–7. https://doi.org/10.1038/277665a0.

Schnitzer MJ, Block SM. Kinesin hydrolyses one ATP per 8-nm step. Nature 1997;388:386–90. https://doi.org/10.1038/41111.

Shelby RD, Hahn KM, Sullivan KF. Dynamic elastic behavior of alpha-satellite DNA domains visualized in situ in living human cells. J Cell Biol 1996;135:545–57. https://doi.org/10.1083/jcb.135.3.545.

Shin GJ-E. Towards a mechanistic understanding of axon transport and endocytic changes underlying paclitaxel-induced peripheral neuropathy. Exp Neurol 2023;359:114258. https://doi.org/10.1016/j.expneurol.2022.114258.

Slayton RE, Park RC, Silverberg SG, Shingleton H, Creasman WT, Blessing JA. Vincristine, dactinomycin, and cyclophosphamide in the treatment of malignant germ cell tumors of the ovary. A Gynecologic Oncology Group Study (a final report). Cancer 1985;56:243–8.

Smoter M, Bodnar L, Duchnowska R, Stec R, Grala B, Szczylik C. The role of Tau protein in resistance to paclitaxel. Cancer Chemother Pharmacol 2011;68:553–7. https://doi.org/10.1007/s00280-011-1696-7.

Srivastava RK, Sasaki CY, Hardwick JM, Longo DL. Bcl-2-mediated drug resistance: inhibition of apoptosis by blocking nuclear factor of activated T lymphocytes (NFAT)-induced Fas ligand transcription. J Exp Med 1999;190:253–65. https://doi.org/10.1084/jem.190.2.253.

Su D, Smith SM, Preti M, Schwartz P, Rutherford TJ, Menato G, et al. Stathmin and tubulin expression and survival of ovarian cancer patients receiving platinum treatment with and without paclitaxel. Cancer 2009;115:2453–63. https://doi.org/10.1002/cncr.24282.

Súarez-Zaizar A, Cárdenas-Cárdenas E, Barajas-Castro YA, Cortés-Esteban P. Controversies on the treatment of ovarian cancer with dose-dense chemotherapy. Chin Clin Oncol 2020;9:53. https://doi.org/10.21037/cco-2019-oc-12.

Sun X, Li D, Yang Y, Ren Y, Li J, Wang Z, et al. Microtubule-binding protein CLIP-170 is a mediator of paclitaxel sensitivity. J Pathol 2012;226:666–73. https://doi.org/10.1002/path.3026.

Sweeney HL, Holzbaur ELF. Motor Proteins. Cold Spring Harb Perspect Biol 2018;10:a021931. https://doi.org/10.1101/cshperspect.a021931.

Tagay Y, Kheirabadi S, Ataie Z, Singh RK, Prince O, Nguyen A, et al. Dynein-Powered Cell Locomotion Guides Metastasis of Breast Cancer. Adv Sci Weinh Baden-Wurtt Ger 2023:e2302229. https://doi.org/10.1002/advs.202302229.

Teneriello MG, Tseng PC, Crozier M, Encarnacion C, Hancock K, Messing MJ, et al. Phase II evaluation of nanoparticle albumin-bound paclitaxel in platinum-sensitive patients with recurrent ovarian, peritoneal, or fallopian tube cancer. J Clin Oncol Off J Am Soc Clin Oncol 2009;27:1426–31. https://doi.org/10.1200/JCO.2008.18.9548.

Tymon-Rosario J, Adjei NN, Roque DM, Santin AD. Microtubule-Interfering Drugs: Current and Future Roles in Epithelial Ovarian Cancer Treatment. Cancers 2021;13:6239. https://doi.org/10.3390/cancers13246239.

Vale RD, Reese TS, Sheetz MP. Identification of a novel force-generating protein, kinesin, involved in microtubule-based motility. Cell 1985;42:39–50. https://doi.org/10.1016/s0092-8674(85)80099-4.

Varga V, Helenius J, Tanaka K, Hyman AA, Tanaka TU, Howard J. Yeast kinesin-8 depolymerizes microtubules in a length-dependent manner. Nat Cell Biol 2006;8:957–62. https://doi.org/10.1038/ncb1462.

Vasey PA, Jayson GC, Gordon A, Gabra H, Coleman R, Atkinson R, et al. Phase III randomized trial of docetaxel-carboplatin versus paclitaxel-carboplatin as first-line chemotherapy for ovarian carcinoma. J Natl Cancer Inst 2004;96:1682–91. https://doi.org/10.1093/jnci/djh323.

Vasquez RJ, Gard DL, Cassimeris L. XMAP from Xenopus eggs promotes rapid plus end assembly of microtubules and rapid microtubule polymer turnover. J Cell Biol 1994;127:985–93. https://doi.org/10.1083/jcb.127.4.985.

Verdier-Pinard P, Pasquier E, Xiao H, Burd B, Villard C, Lafitte D, et al. Tubulin proteomics: towards breaking the code. Anal Biochem 2009;384:197–206. https://doi.org/10.1016/j.ab.2008.09.020.

Vergote I, Coens C, Nankivell M, Kristensen GB, Parmar MKB, Ehlen T, et al. Neoadjuvant chemotherapy versus debulking surgery in advanced tubo-ovarian cancers: pooled analysis of individual patient data from the EORTC 55971 and CHORUS trials. Lancet Oncol 2018;19:1680–7. https://doi.org/10.1016/S1470-2045(18)30566-7.

Vergote I, Tropé CG, Amant F, Kristensen GB, Ehlen T, Johnson N, et al. Neoadjuvant chemotherapy or primary surgery in stage IIIC or IV ovarian cancer. N Engl J Med 2010;363:943–53. https://doi.org/10.1056/NEJMoa0908806.

Verweij J, Clavel M, Chevalier B. Paclitaxel (Taxol) and docetaxel (Taxotere): not simply two of a kind. Ann Oncol Off J Eur Soc Med Oncol 1994;5:495–505. https://doi.org/10.1093/oxfordjournals.annonc.a058903.

Walczak CE, Cai S, Khodjakov A. Mechanisms of chromosome behaviour during mitosis. Nat Rev Mol Cell Biol 2010;11:91–102. https://doi.org/10.1038/nrm2832.

Walker JL, Brady MF, Wenzel L, Fleming GF, Huang HQ, DiSilvestro PA, et al. Randomized Trial of Intravenous Versus Intraperitoneal Chemotherapy Plus Bevacizumab in Advanced Ovarian Carcinoma: An NRG Oncology/Gynecologic Oncology Group Study. J Clin Oncol Off J Am Soc Clin Oncol 2019;37:1380–90. https://doi.org/10.1200/JCO.18.01568.

Wani MC, Taylor HL, Wall ME, Coggon P, McPhail AT. Plant antitumor agents. VI. The isolation and structure of taxol, a novel antileukemic and antitumor agent from Taxus brevifolia. J Am Chem Soc 1971;93:2325–7. https://doi.org/10.1021/ja00738a045.

Wegner A. Head to tail polymerization of actin. J Mol Biol 1976;108:139–50. https://doi.org/10.1016/s0022-2836(76)80100-3.

Williams SD, Blessing JA, Moore DH, Homesley HD, Adcock L. Cisplatin, vinblastine, and bleomycin in advanced and recurrent ovarian germ-cell tumors. A trial of the Gynecologic Oncology Group. Ann Intern Med 1989;111:22–7. https://doi.org/10.7326/0003-4819-111-1-22.

Wu H, Duan Y, Gong S, Zhu Q, Liu X, Liu Z. An Integrative Pan-Cancer Analysis of Kinesin Family Member C1 (KIFC1) in Human Tumors. Biomedicines 2022;10:637. https://doi.org/10.3390/biomedicines10030637.

Yuan J, Lan H, Jiang X, Zeng D, Xiao S. Bcl-2 family: Novel insight into individualized therapy for ovarian cancer (Review). Int J Mol Med 2020;46:1255–65. https://doi.org/10.3892/ijmm.2020.4689.

Yusuf RZ, Duan Z, Lamendola DE, Penson RT, Seiden MV. Paclitaxel resistance: molecular mechanisms and pharmacologic manipulation. Curr Cancer Drug Targets 2003;3:1–19. https://doi.org/10.2174/1568009033333754.

Zhang B, Liu P, Li Y, Hu Q, Li H, Pang X, et al. Multi-omics analysis of kinesin family member 2C in human tumors: novel prognostic biomarker and tumor microenvironment regulator. Am J Cancer Res 2022;12:4954–76.

Zhang CC, Yang JM, White E, Murphy M, Levine A, Hait WN. The role of MAP4 expression in the sensitivity to paclitaxel and resistance to vinca alkaloids in p53 mutant cells. Oncogene 1998;16:1617–24. https://doi.org/10.1038/sj.onc.1201658.

Zhang R, Alushin GM, Brown A, Nogales E. Mechanistic Origin of Microtubule Dynamic Instability and Its Modulation by EB Proteins. Cell 2015;162:849–59. https://doi.org/10.1016/j.cell.2015.07.012.

# Tubulin Complexity in Cancer and Metastasis

# 2

Michael Danziger, Fuhua Xu, Helen Noble, Peixin Yang, and Dana M. Roque

**Abstract**

Tubulin plays a fundamental role in cellular function and as the subject for microtubule-active agents in the treatment of ovarian cancer. Microtubule-binding proteins (e.g., tau, MAP1/2/4, EB1, CLIP, TOG, survivin, stathmin) and posttranslational modifications (e.g., tyrosination, deglutamylation, acetylation, glycation, phosphorylation, polyamination) further diversify tubulin functionality and may permit additional opportunities to understand microtubule behavior in disease and to develop microtubule-modifying approaches to combat ovarian cancer. Tubulin-based structures that project from suspended ovarian cancer cells known as *microtentacles* may contribute to metastatic potential of ovarian cancer cells and could represent an exciting novel therapeutic target.

M. Danziger · H. Noble · P. Yang
Department of Obstetrics, Gynecology, and Reproductive Sciences, University of Maryland School of Medicine, Baltimore, MD, USA

F. Xu · D. M. Roque (✉)
Division of Gynecologic Oncology, Greenebaum Comprehensive Cancer Center, University of Maryland School of Medicine, Baltimore, MD, USA
e-mail: droque@som.umaryland.edu

**Keywords**

Tubulin code · Posttranslational modifications · Microtentacle · Detyrosination · Polyglutamylation · Acetylation · Polyglycylation · Tau · EB1 · Microtubule-associated protein

## 2.1 Tubulin Complexity

In the preceding chapter, we introduced the fundamental role of tubulin in cellular function and of microtubule-active agents in the treatment of ovarian cancer. Nevertheless, some tumors continue to evade standard agents; therefore, there lies interest in the establishment of novel microtubule-targeting agents, including various taxoids (e.g., larotaxel [XRP9881, Sanofi], BMS-184476, TPI-287), dolastatins (e.g., dolastatin 10, auristatin, cemadotin, tasidotin), colchicine analogs (e.g., ZD6126, combretastatin A-4, 2-methoxyestradiol, mivobulin isethionate), sulfonamides (e.g., ABT-751, HMN-214), sea sponge derivatives (e.g., halichondrin B, eribulin mesylate, hamiasterlin, discodermolide, peloruside, ceratamines, LY573636), algae or fungal derivatives (e.g., cryptophycin 52, rhizoxin), and maytansinoids (e.g., cantuzumab mertansine, huN901-DM1/BB-10901), among others. Currently, all agents approved by the Food and



H. Schatten (ed.), *Cell and Molecular Biology of Ovarian Cancer*, Advances in Experimental Medicine and Biology 1452, https://doi.org/10.1007/978-3-031-58311-7_2

Drug Administration bind β-tubulin. β-tubulin binding sites for pharmacologic agents include the taxane site (along the luminal pocket), laulimalide/peloruside site (at the lateral interface between tubulin protofilaments), colchicine site (along the intermediate domain near the α/β interface), vinca site (between the longitudinal interface of protofilaments that prevent conformational changes), and maytansine site (at the exposed side of the β-tubulin subunit) (Eli et al. 2022). Pironetin is the only known microtubule-inhibiting agent known to bind α-tubulin, using a site located at the α-/β-tubulin interface. Microtubule-binding proteins and posttranslational modifications further diversify tubulin functionality and may permit additional opportunities to understand microtubule behavior in disease and develop microtubule-modifying approaches to combat ovarian cancer.

## 2.2 Microtubule-Associated Proteins (MAPs)

As evidenced by their stochastic rapid dynamics, microtubules are relatively unstable yet central to many coordinated and non-random cellular processes. In vivo, MAPs determine microtubule stability and function. MAPs have either microtubule stabilizing or destabilizing effects, which they achieve by activity either at the microtubule ends or within the lattice itself.

MAPs also have clinical significance, as many are overexpressed in a wide variety of cancers, including ovarian. This qualifies them as potential therapeutic targets for these cancers. Given their ubiquity, targeting them may prove to be a problematic strategy; however, as tumor genotyping becomes more accessible, manipulating specific MAPs may become a reasonable approach. While no means an exhaustive summary, we discuss below some of the best characterized MAPs.

***Tau*** Tau (tubulin-associated unit) consists of six isoforms generated from alternative splicing from mRNA of the *MAPT* gene on chromosome 17q21. The primary transcript contains 16 exons. Exons 2, 3, and 10 are alternatively spliced and specific to human adult brain. Tau, along with MAP2, constitutes the dominant MAP within the central nervous system. Tau, MAP2, and MAP4 are structurally related, through a common C-terminal KXGS motif that can be phosphorylated (Illenberger et al. 1996). MAP4 is primarily non-neuronal. Tau mRNA has been detected in heart, lung, pancreas, kidney, muscle, and testis (Dehmelt and Halpain 2005).

Tau acts as a microtubule stabilizer, particularly of axonal microtubules. The tau C-terminus contains repetitive regions (R1-R4) encoded by exons 9–12 that bind microtubules (Dehmelt and Halpain 2005; Robert and Mathuranath 2007; Barbier et al. 2019). Two categories of C-terminal repeats can result from alternative splicing: a more potent microtubule-binding 4-repeat form and a less-potent 3-repeat form. The binding of Tau to microtubules stabilizes them, and thereby prevents depolymerization and catastrophe (Cassimeris and Spittle 2001; Goodson and Jonasson 2018). Overexpression of tau leads to the accumulation of microtubules, increased microtubule stability, and reversal of cell flexibility induced by ovarian cancer drugs such as paclitaxel (Dye et al. 1993).

Tau activity and affinity to microtubules are regulated by phosphorylation of the KXGS motif within the microtubule-binding repeat by PKA and MARK family kinases (Ozer and Halpain 2000; Dehmelt and Halpain 2005; Illenberger et al. 1996). Phosphorylation of tau decreases microtubule-binding affinity. Hyperphosphorylated tau loses its ability to attach to microtubules and is secreted into the extracellular space, where it self-aggregates into neurofibrillary tangles.

Despite the strong localization of tau to neurons, it has been shown to be a potential marker of response to paclitaxel in breast and ovarian cancers, and tau-negative tumor status has been shown to confer improved response to paclitaxel (Barbolina 2022; Smoter et al. 2013); however, studies regarding the prognostic value of tau on overall (OS) and progression-free (PFS) survival remain conflicting (Steffensen et al. 2014). It is hypothesized that tau can compete with taxanes for the microtubule-binding site (Downing and Nogales 1999). Silencing of tau by siRNA sensitizes ovarian carcinoma cells to paclitaxel in a collagen matrix (Gurler et al. 2015). Because of this, tau modulation as an adjunct to microtubule-targeting agents is an attractive strategy (Yamauchi et al. 2017).

***MAP 1*** The MAP1 family includes *MAP1A* on chromosome 15q15.3 and *MAP1B* on 5q13. Both are synthesized as a precursor polypeptide that is processed by proteolytic cleavage to yield heavy chains (MAP1A-HC, 350 kDa; MAP1B-HC, 300 kDa) and light chains (MAP1A-LC2, 28 kDa; MAP1B-LC1, 34 kDa). LC3, a third light chain, is a subunit of both MAP1A and MAP1B that plays an important role in autophagy (Tanida et al. 2008). The MAP1 family promotes microtubule stability by preventing depolymerization (Vandecandelaere et al. 1996). MAP1B has weaker microtubule-stabilizing capability compared to MAP2 and produces wavy microtubule bundles versus the straight, rigid bundles produced by overexpression of MAP2 and tau (Noiges et al. 2002; Felgner et al. 1997). In some human cancers, patients with high levels of p53 and LC3A co-expression have poor clinical prognoses (Wang et al. 2013).

***MAP 2*** MAP2 is a neuronal-specific protein encoded by the *MAP2* gene on chromosome 2q34. Like tau and MAP4, MAP2 contains a KXGS motif that can be phosphorylated to decrease microtubule affinity. At least four isoforms generated by alternative splicing exist, the most important of which appear to be MAP2a, MAP2b, and MAP2c (Dehmelt and Halpain 2004, 2005). MAP2c is the juvenile isoform and a weak stabilizer of microtubules; MAP2a is expressed when MAPc levels begin to decline, and MAP2b is expressed during development and adulthood (Garner et al. 1988; Chung et al. 1996). Like tau, MAP2 isoforms bind microtubules longitudinally and stabilize microtubules by preventing catastrophe (Gamblin et al. 1996; Al-Bassam et al. 2002). Interestingly, MAP2 clusters around microtubules and may serve as physical impediment to catastrophe (Ichihara et al. 2001).

Variable expression of MAP2 has been demonstrated in a number of cancers (Liu et al. 2001; Fang et al. 2001). While dysregulated expression has not been described as prognostic in ovarian cancer, increased expression of MAP2 has been described in taxane-sensitive pancreatic ductal carcinoma compared to resistant disease (Veitia et al. 2000). Additionally, vincristine-resistant neuroblastoma cells display a marked decrease in MAP2c without major changes in other isoforms, suggesting that vincristine resistance in these cells can result from a shift in isoform equilibrium to the stronger stabilizing isoforms MAP2a and MAP2b, resulting in hyperstable microtubules (Don et al. 2004).

***MAP 4*** MAP4 is the major MAP found in non-neuronal cells. *MAP4* is located on chromosome 3p21 and encodes a heterogeneous polypeptide that undergoes posttranslational proteolysis. Like tau and MAP2, MAP4 stabilizes microtubules by binding the microtubule lattice and slows shrinkage and catastrophe. MAP4-related microtubule stability has been implicated in intracellular trafficking of organelles and spindle fiber formation (Bulinski and Borisy 1980; Parysek et al. 1984a, b). Overexpression of MAP4 has been shown to both stabilize microtubules and inhibit microtubule functions reliant upon dynamic instability, such as vesicular transport and cellular motility (Bulinski et al. 1997).

Unlike tau and MAP2, MAP4 has been shown to enhance microtubule polymerization, sensitivity to paclitaxel, and paclitaxel to the microtubule (Zhang et al. 1998). MAP4 phosphorylation and subsequent dissociation from microtubules is associated with decreased sensitivity to paclitaxel, while vinblastine-resistant cells produce non-phosphorylated forms (Poruchynsky et al. 2001; Kavallaris et al. 2001). MAP4 expression is also negatively regulated by p53, and induction of p53 leads to *MAP4* repression and reduced sensitivity to paclitaxel but increased sensitivity to vinblastine (Murphy et al. 1996; Zhang et al. 1998; Kavallaris 2010). By contrast, increased expression of MAP4 and class III β-tubulin seems to correlate with resistance of human leukemia cells to a microtubule-targeting desoxyepothilone B (Verrills et al. 2003).

***EB1*** EB1 (end-binding protein 1), encoded by the *MAPRE1* gene on chromosome 20q11.21, is a member of the evolutionarily conserved family of end-binding microtubule plus end tracking proteins (+TIP) that localize to and track with the growing end (Mimori-Kiyosue et al. 2000; Akhmanova and Steinmetz 2010). It is one of three EB proteins (including EB2 and EB3) in mammals. This family shares a conserved N-terminal domain that adopts a calponin homology (CH) fold that binds microtubules (Hayashi and Ikura 2003). EB1 was first identified as a binding partner of APC (adenomatous polyposis coli, a tumor suppressor involved in hereditary colon cancer) (Schuyler and Pellman 2001; Su et al. 2009). EB family proteins are considered to be master regulators of +TIP complexes because they recognize growing microtubule ends and recruit other +TIPs (including CLIP170 and $p150^{Glued}$) to them (Honnappa et al. 2009; Akhmanova and Steinmetz 2015). EB1 possesses an intrinsic ability to promote microtubule assembly and prevent pauses in growth (Ligon et al. 2003; Komarova et al. 2009). EB1 accumulates at the end of microtubules where it forms "comet-like" aggregations that disappear during shrinkage, affirming the role of EB1 in microtubule growth (Matov et al. 2010; Bieling et al. 2007; Piehl and Cassimeris 2003). Furthermore, EB1 organizes other +TIPs to form +TIP complexes at the plus end. It does this through two domains: an EEY/F motif at the C-terminus which interacts with CAP-Gly domain-containing proteins (e.g., CLIPs) or an EB homology (EBH) domain which is associated with the SxIP motif present on other +TIPs. The EEY/F motif mimics tyrosinated α-tubulin (Perez et al. 1999; Kumar and Wittmann 2012, p. 2).

Breast cancers have been shown to upregulate EB1, though data regarding the effects of EB1 on paclitaxel sensitivity remain conflicting (Dong et al. 2010; Luo et al. 2014; Thomas et al. 2015). EB1 overexpression may enhance the effects of vinca alkaloids (Berges et al. 2014). Upregulation of EB1 was found to be associated with increased OS in ovarian cancer, which would support a model in which EB1 expression increases paclitaxel sensitivity (Schiewek et al. 2018). Additional in vivo characterization of the relationship between EB1 and microtubule-active agents in ovarian cancer is required to solidify its potential as a marker for chemotherapeutic sensitivity.

***CLIP Family*** Cytoplasmic linker protein 170 (CLIP170, also known as CLIP1), encoded by the *CLIP1* gene on chromosome 12q24.31, was the first MAP identified as a +TIP (Rickard and Kreis 1990; Perez et al. 1999; Pierre et al. 1992). It contains an N-terminal CAP-gly (cytoskeleton-associated protein-glycine-rich) domain and tandem C-terminal zinc knuckle motifs as well as an EEY/F motif. The CAP-gly domain mediates both microtubule and EB protein binding through interaction with the EEY/F motif on EB1 (Weisbrich et al. 2007; Honnappa et al. 2006). The CAP-gly domain is also present on $p150^{Glued}$, a subunit of dynactin, a dynein activator complex (Galjart 2005). Like many +TIPs, CLIP170 also contains an SxIP motif, which is necessary and sufficient to target it to growing microtubule ends and functions as a microtubule tip-localization signal (Honnappa et al. 2009). The plus end tracking of CLIP170 is mediated by treadmilling: individual CLIP170 proteins associate with the newly polymerized distal end and dissociate from the older lattice while remaining immobile with respect to the microtubule (Perez et al. 1999). CLIP170 has been shown to both stabilize microtubules and counteract catastrophe (Carvalho et al. 2004).

CLASPs (CLIP-associating proteins) interact with CLIPs at the microtubule plus end and suppress catastrophe and promote rescue by recruiting tubulin dimers to the plus end (Galjart 2005; Al-Bassam et al. 2010). CLASPs are also involved in helping CLIPs coordinate other microtubule activities, including anchoring microtubules to the cell cortex through interaction with LL5β and helping coordinate microtubule binding to the kinetochore (Maiato et al. 2004; Lansbergen et al. 2006).

The expression of CLIP170 correlates with paclitaxel sensitivity in breast cancer cell lines and sensitivity to paclitaxel-based chemotherapy (Sun et al. 2012; Xie et al. 2016). A relationship between CLIP170 and taxane response would be expected given the relationship between CLIP170 and EB1. As with EB1, these data make +TIPs a focus for anti-cancer therapy.

***TOG family*** TOG (tumor overexpressed gene) family proteins are found across diverse species and characterized by up to five TOG domains at the N-terminus. Each TOG domain encompasses up to six HEAT repeats, a protein motif characterized by two linked α helices that forms a flat, paddle-like structure that binds tubulin, among other substrates (Al-Bassam et al. 2007). The TOG family is highly conserved and has been found throughout eukaryotes. They can bind directly to microtubules and have an intrinsic ability to promote microtubule polymerization at the plus end. As a result, they belong to the large family of +TIPs. CKAP5 (formerly known as TOGp) is the TOG family protein in humans, encoded by the *CKAP5* gene on chromosome 11p11.2. TOG function has been extensively studied in XMAP215, the *Xenopus* homolog of CKAP5. Purified XMAP215 added to microtubules increases the plus end elongation rate by approximately tenfold but only about twofold at the minus end (Vasquez et al. 1994). In contrast to other MAPs, this activity is partially mediated by the intrinsic polymerase activity of CKAP5, which allows it to catalyze the addition of up to 25 tubulin dimers to a microtubule (Brouhard et al. 2008). More recent studies have demonstrated that the plus end stabilizing activity of XMAP215 can also counterbalance the catastrophe-promoting activity of XKCM1 (the *Xenopus* homolog of MCAK, a microtubule depolymerizing kinesin-13), and depletion of MCAK has been shown to result in longer microtubules, while mutation of *CKAP5* results in shorter spindles (Tournebize et al. 2000; Domnitz et al. 2012; Cassimeris and Morabito 2004). Disruptions in TACC1 and TACC3, which complex with CKAP5, have been described in ovarian cancer (Lauffart et al. 2005).

***Survivin*** Survivin is a member of the inhibitor of apoptosis (IAP) family. It both suppresses apoptosis and regulates cell division. It is encoded by *BIRC5* on chromosome 17q25.3. It localizes to both spindle microtubules and kinetochore microtubules, promoting microtubule stability (Rosa et al. 2006). Survivin binds to centromeres on a para-polar axis during prophase/metaphase, relocates to the spindle midzone during anaphase/telophase, and disappears at the end of telophase (Cheung et al. 2009). It functions as an oncogene and is overexpressed in nearly every human tumor. It aids tumorigenesis by promoting evasion of apoptosis and dysregulation of cell division. Its overexpression is associated with treatment resistance and poor outcome (Altieri 2003). Wildtype p53 suppresses expression of survivin at both the mRNA and protein levels by binding the *BIRC5* promoter. Thus, loss of function in *TP53* will result in loss of negative regulation of p53 on survivin (Raj et al. 2008). Indeed, most ovarian cancers show survivin expression, which correlates with poor prognostic markers (high grade, histologic type, p53 mutation) (Cohen et al. 2003).

***Stathmin*** Stathmin (Op18 or oncoprotein 18) is a microtubule-destabilizing protein encoded by the *STMN1* gene on chromosome 1q36. It was discovered as an oncoprotein that is highly expressed in leukemia, breast, and ovarian cancers. It is a member within a family of structurally related destabilizing proteins that includes SCG10 and SCLIP (Rubin and Atweh 2004; Cassimeris 2002). There are two known mechanisms by which stathmin destabilizes microtubules: (1) sequestration of tubulin heterodimers by directly binding two dimers in a curved configuration that prevents incorporation into the growing microtubule—this promotes depolymerization by limiting the available substrate for incorporation into the growing end of the microtubule, ultimately leading to loss of the GTP cap and subsequent catastrophe and (2) interference with lateral bonds between tubulin heterodimers, with resultant destabilization of the microtubule tip and catastrophe (Gupta et al. 2013).

Phosphorylation of stathmin leads to loss of functionality (Amayed et al. 2002). It is found mostly in the phosphorylated state in cells undergoing mitosis and is more likely to be dephosphorylated in interphase cells. Paclitaxel has been shown to promote phosphorylation at Ser-16, thereby serving as a negative regulator of stathmin function (Liu et al. 2013). *STMN1* overexpression is associated with paclitaxel resistance and poor prognosis in ovarian cancer (Su et al. 2009). This is perhaps unsurprising given the antagonistic functionalities of paclitaxel and stathmin: taxanes induce microtubule stabilization/polymerization while stathmin promotes destabilization/depolymerization. Overexpression of stathmin results in significantly decreased paclitaxel binding to microtubules, providing a potential mechanism of taxane resistance in ovarian cancers (Alli et al. 2002). Intriguingly, the anti-cancer effects of the vinca alkaloid vinblastine appears to be negatively impacted by stathmin overexpression (Alli et al. 2002).

## 2.3 Tubulin Posttranslational Modifications (PTMs)

Tubulin-based processes such as cell growth, differentiation, motility, and intracellular trafficking (Avila 1992) all require proper posttranslational modifications of microtubule proteins. Thus, understanding tubulin PTMs remains essential for realizing the role of microtubules in both the normal and pathologic states.

PTMs are highly dynamic and often reversible processes by which a protein's functional properties are altered either via the covalent linkage to a chemical group or polymer or via the removal of amino acid residues from the peptide chain. Tubulin PTMs usually occur post-polymerization, preferentially on the α-/β-tubulin heterodimers of stable, rather than dynamic, microtubules. Several enzymes, which are either ubiquitous or unique to tubulin, can catalyze the PTM processes. PTMs take place on the C-terminal of the tubulin peptide and include polyglutamylation, polyglycylation, acetylation, phosphorylation, and detyrosination (Barra et al. 1973a; Barra et al. 1974). PTMs mark subpopulations of microtubules and selectively affect downstream microtubule-based functions (Raybin and Flavin 1977). During this process, modifications on the tubulin peptide generate a "tubulin code" specified by the nature, length, and spacing patterns of these modifications that can be read by MAPs in a manner analogous to the histone code directing diverse chromatin functions (Raybin and Flavin 1977). The most common PTMs are summarized below.

***Detyrosination*** In 1973, Barra and colleagues first described α-tubulin as the substrate for detyrosination and tyrosination modification processes (Barra et al. 1973a, 1974). The α-tubulin gene encodes a tyrosine residue at the C-terminus of the peptide chain. The initial detyrosination is catalyzed by tubulin carboxypeptidase (TCP) and results in Glu-α-tubulin, which has an exposed glutamate residue at the newly formed C-terminus. Tyrosination is catalyzed by tubulin-tyrosine ligase (TTL), the first enzyme identified as a tubulin-modifying enzyme (Barra et al. 1973b). Detyrosination and tyrosination are reversible PTMs.

The tubulin detyrosination-tyrosination cycle remains elusive. Tubulin detyrosination is thought to evoke microtubule stability: long-lived microtubules in various cells are noted to lack tyrosination, although detyrosination itself does not lead to microtubule stabilization (Webster et al. 1990). In addition, it has been shown that tubulin detyrosination plays an important role in the maintenance of mitotic spindle size and accurate spindle position during cell division (Liao et al. 2019). The clinical impact of the tubulin detyrosination-tyrosination cycle remains under investigation as TTL is downregulated in cancer cells of animals and humans. Indeed, low levels of TTL and, correspondingly, tyrosinated tubulin correlate with an increased risk of tumor development, invasiveness (Miller et al. 2008), and paclitaxel resistance (Banerjee 2002).

***Polyglutamylation*** Among the known tubulin PTMs, polyglutamylation is the most recently discovered process involving both α and β tubulin. Polyglutamylation involves the addition of one or more glutamate residues to the gamma-carboxyl group of glutamate residues in the peptide chain. The process generates glutamate side chains of variable length on the α-/β-tubulin heterodimers (Janke et al. 2005, 2008). Polyglutamylation is catalyzed by members of the tubulin tyrosine ligase-like (TTLL) family, each of which has a particular reaction preference, such as α-/β-tubulin specific or glutamate side chain initiation/elongation specific (Janke et al. 2005). Polyglutamylation modification can be reversed (deglutamylation) by enzymes that belong to the cytosolic carboxypeptidase (CCP) family (Rogowski et al. 2010). Similar to the TTL enzyme specificity, some CCPs preferentially shorten long glutamate chains (e.g., CCP1, CCP4, and CCP6), whereas others remove the branching point of side chains (e.g., CCP5) (Rogowski et al. 2010).

Polyglutamylation and deglutamylation may help stabilize and destabilize microtubules, respectively. These modification processes can also affect the interaction of microtubules with kinesins, MAPs, and microtubule-severing factors through altered binding affinity determined by the glutamate chain length and location (Goodson and Jonasson 2018). Microtubules that comprise the mitotic spindle, neuronal projections, centrioles/basal bodies, and axenomes undergo tubulin polyglutamylation. Tubulin polyglutamylation levels are temporarily increased in the central mitotic spindle and the midbody during cell division (Bobinnec et al. 1998). The complexity and distribution of glutamylated tubulin isoforms follow a unique pattern in the mouse sperm flagellum (Kann et al. 2003). Microtubules of the flagellum contain polyglutamylated isoforms except in the tip where only monoglutamylated isoforms are detected. Using breast cancer as a model of disease in which mechanical signaling promotes progression, tissue matrix stiffening alters glutamine metabolism to promote tubulin polyglutamylation and to enforce microtubule stabilization, thereby promoting invasive potential (Arnold et al. 2020). Pharmacologic inhibition of glutamine metabolism decreases tubulin polyglutamylation and the mechanical stabilization of microtubules. Similarly, limiting tubulin polyglutamylation by overexpressing mutated tubulin lacking glutamylation sites decreases microtubule stability, thereby hampering cancer aggressiveness in vitro and in vivo (Torrino et al. 2021).

***Acetylation*** Acetylation may be the most well-studied tubulin PTM. The peculiarity of tubulin acetylation is that the modification sites are located at the luminal surface of microtubules (Hammond et al. 2008). Acetylation of the epsilon amino group of lysine at position 40 (Lys40 or K40) occurs on α-tubulin of assembled microtubules, including those of cytoplasmic, spindle, centriolar, and axonemal origins (Hammond et al. 2008). K40 acetylation is catalyzed by α-tubulin acetyltransferases (αTAT) and MEC-17 (Szyk et al. 2014; Akella et al. 2010; Davenport et al. 2014). Another acetylation site occurs at K252 of β-tubulin (Chu et al. 2011). Acetylation of K252 is catalyzed by the San acetyltransferase and takes place on soluble tubulin heterodimers instead of tubulins in microtubules. K252 of β-tubulin localizes at the interface of tubulin heterodimers and interacts with the phosphate group of the α-tubulin-bound GTP. K252 acetylation slows tubulin incorporation into microtubules by neutralizing the positive charge on K252 and allowing tubulin heterodimers to adopt a conformation that disfavors tubulin incorporation (Chu et al. 2011). Tubulin acetylation enhances resistance to paclitaxel-induced cell death through Mcl-1 stabilization (Wattanathamsan et al. 2021).

Similar to detyrosination, tubulin acetylation is considered a marker of microtubule age. A high degree of acetylation indicates stable microtubules resistant to depolymerizing conditions, although acetylation itself does not lead to microtubule stabilization (Palazzo et al. 2003). Deacetylated microtubules break under mechanical stress and are less flexible than acetylated microtubules (Xu et al. 2017). Cryo-electron microscopy has shown that tubulin acetylation induces electrostatic perturbations that restrict the motion of the αK40 loop and, therefore, weakens the lateral contacts and reduces inter-protofilament interactions. The increased flexibility results from fewer inter-protofilament interactions; as such, αTAT may locally fine-tune the load-bearing capacity and mechanical resistance of microtubules (Eshun-Wilson et al. 2019).

***Polyglycylation*** Polyglycylation occurs on both α- and β-tubulins. Modifications on multiple sites in the C-terminal of the peptide create side chains of glycine with various lengths (Redeker et al. 1994). The enzymes that catalyze tubulin polyglycylation (glycylases) are members of the TTLL family (Janke et al. 2005). In mammals, these enzymes can be subdivided into initiating and side-chain-elongating glycylases that are specific to α- or β-tubulin (Bosch Grau et al. 2013). In contrast, glycylases in *Drosophila* are multifunctional enzymes facilitating both initiation and elongation processes for both tubulin isoforms (Janke et al. 2005).

Tubulin polyglycylation takes place commonly in microtubules of the axoneme. A *Ttll*-knockout mouse model demonstrated the importance of tubulin polyglycylation in controlled flagellar beating, directional sperm swimming, and male fertility (Gadadhar et al. 2021). Glycylation accumulates in primary cilia in a length-dependent manner, which strongly suggests that glycylation is essential for the homeostasis of primary cilia (Gadadhar et al. 2017). Glycylation may play a role in recruitment of microtubule-severing proteins (Roll-Mecak and McNally 2010). Abnormal formation or function of cilia has been linked to a broad range of human genetic disorders termed ciliopathies (Waters and Beales 2011). *Ttll3* knockout results in reduced primary cilia, increased cell proliferation, and promotion of colon cancer (Rocha et al. 2014). Clinically, TTLL levels are associated with the FIGO (Fédération Internationale de Gynécologie et Obstétrique) stage ($P = 0.001$) and peritoneal cytology ($P = 0.042$); TTLL12 is thought to be an independent risk factor for OS ($p = 0.022$) and disease-free survival ($p = 0.040$) in ovarian cancer (Yang et al. 2020).

***Phosphorylation*** Phosphorylation of tubulin was discovered in the early 1970s (Goodman et al. 1970). While most tubulin PTMs act on tubulin subunits that are already incorporated into microtubules, phosphorylation can take place on both the α-/β-tubulin heterodimers and polymers at serine, threonine, and tyrosine residues (Wloga et al. 2017). To date, only one precise tubulin phosphorylation site has been functionally analyzed, which is Serine172 (Ser172) (Fourest-Lieuvin et al. 2006). Ser172 phosphorylation only occurs in β-tubulin. Other serine and tyrosine residues on the α/β tubulin heterodimers and microtubules are presumed targets of phosphorylation, with unknown function to date (Ma and Sayeski 2007). A variety of kinases have been reported to be critical for tubulin phosphorylation (Gard and Kirschner 1985; Kadowaki et al. 1985; Fernandez et al. 1999). Cyclin-dependent kinase 1 (CDK1) catalyzes Ser172 phosphorylation in β-tubulin of the α-/β--tubulin heterodimers in vivo and in vitro (Fourest-Lieuvin et al. 2006). Spleen tyrosine kinase (SYK), a non-receptor tyrosine kinase, specifically targets tyrosine residues in α-tubulin. Phosphorylated α-tubulin can then be assembled into microtubules in vitro. SYK-dependent tubulin phosphorylation may affect the ability of the microtubule cytoskeleton to serve as a platform upon which signaling complexes are assembled (Faruki et al. 2000). The phosphorylation state of Ser165 in α-tubulin causes breast cells to exhibit hyper-proliferation (Markovsky et al. 2018).

An early study suggested that phosphorylation of β-tubulin was a key step to regulate microtubule assembly and function during cell differentiation (Gard and Kirschner 1985). Tubulin phosphorylation has also been found to play an essential role in regulating mitosis, likely by controlling microtubule assembly and number (Fourest-Lieuvin et al. 2006; Ferreira et al. 2018). The physiological relevance of tubulin phosphorylation remains an active area of research with various studies supporting its function in regulating microtubule polymerization either positively or negatively.

***Polyamination*** Cold typically induces tubulin depolymerization. A cold-resistant, basic tubulin can be generated by addition of polyamine to the highly conserved glutamine 15 (Q15) of β-tubulin (Brady et al. 1984), which is near the GTP pocket.

***Other*** Tubulin succination has been observed in dynamic microtubules, but the cellular consequences are incompletely described. In this process, S-2-succino cysteine is created through a reaction between fumarate and thiol groups (Alderson et al. 2006). O-Glc-NAcylation has been described at Ser178 and Thr179 of α-tubulin and peptides 216–238 of β-tubulin (Ji et al. 2011).

The cellular functions of tubulin PTMs in ovarian cancer remain incompletely elucidated. The complexity of permutations is virtually infinite, as microtubules can undergo multiple PTMs. Although initial insights have already revealed the potential roles of tubulin PTMs in several human pathologies, such as neurodegeneration, ciliopathy, and cancer, more studies are required to address how this tubulin code allows cells to divide, migrate, communicate, and differentiate in an orderly manner.

## 2.4 Microtentacles

Tubulin-based protrusions termed "microtentacles" have been visualized in floating cancer cells upon their detachment (Whipple et al. 2007). Though first observed in breast cancer, they have since been described in ovarian cancers (Reader 2021). At diagnosis, roughly 80% of ovarian cancer patients exhibit intra-abdominal metastases, often in association with ascites, which may facilitate dissemination of exfoliated cells (Penet et al. 2018). Microtentacles may therefore be particularly important to the pathophysiology of ovarian cancer. They are supported by vimentin (Whipple et al. 2008) and promoted by actin-destabilizing agents, distinguishing them from other classical locomotory cellular structures including lamellipodia, filopodia, invadopodia, and podosomes (Alblazi and Siar 2015). Lamellipodia tend to be broad-based protrusions that lead migrating cells. Their actin organization is orthogonal with branching at the advancing edge. The formation of lamellipodia is tied to epithelial-mesenchymal transition (Khoury et al. 2001). Filopodia are thin and have a parallel actin arrangement (Suraneni et al. 2012). Filopodia and lamellipodia function primarily in two-dimensional space, whereas podosomes and invadopodia are three-dimensional. The latter therefore requires the capacity to degrade matrix. Podosomes are roughly 0.4 micrometers in length and 1 micrometer in width, with columnar actin arrays (David-Pfeuty and Singer 1980). Highly migratory normal cells such as macrophages, dendritic cells, osteoclasts, smooth muscle cells, and endothelial cells display podosomes (Yilmaz and Christofori 2009). Invadopodia are essentially malignant podosomes with invasive capacity, ranging from 0.1 to 0.8 micrometers in diameter with extensions up to 2 micrometers (Artym et al. 2006). Together podosomes and invadopodia constitute the "invadosome" complex (Génot and Gligorijevic 2014). Podosomes and invadopodia are F-actin rich. They are regulated by actin regulators including WASP, NWASP contactin, Src kinase, Arp 2/3, MT1-MMP.

In ovarian cancer cells and freshly procured ascites, the lengths and patterns of microtentacles appear to vary by histology and biologic behavior. For example, clear cell ovarian carcinomas, which tend to be diagnosed more frequently at an early stage (on-metastatic), have fewer and shorter microtentacles compared to high-grade serous carcinomas that almost always present at an advanced stage with intra-abdominal metastases. Clear cell carcinomas are more likely to overexpress class III β-tubulin, but equivalent levels of acetylated α-tubulin. Across multiple histologies, microtubule-depolymerizing drugs decrease microtentacle dynamics, suggesting that they are tubulin-based. Ovarian cancer cell lines such as JHOC-7 that fail to form microtentacles express high levels of p-cofilin and elevated levels of tyrostinated α-tubulin, whereas cell lines that possess microtentacles favor detyrostinated α-tubulin. Microtentacles appear to participate in aggregation of floating cells: application of tubulin-depolymerizing and -stabilizing agents affects clustering capacity in cell lines that form microtentacles but less so in those void of microtentacles (Reader 2021). Microtentacles therefore may participate in key aspects of ovarian cancer progression and cell-cell interactions, and are thus prime subjects for tubulin-based therapies that may disrupt metastasis.

## References

Akella JS, Wloga D, Kim J, Starostina NG, Lyons-Abbott S, Morrissette NS, et al. MEC-17 is an alpha-tubulin acetyltransferase. Nature 2010;467:218–22. https://doi.org/10.1038/nature09324.

Akhmanova A, Steinmetz MO. Control of microtubule organization and dynamics: two ends in the limelight. Nat Rev Mol Cell Biol 2015;16:711–26. https://doi.org/10.1038/nrm4084.

Akhmanova A, Steinmetz MO. Microtubule +TIPs at a glance. J Cell Sci 2010;123:3415–9. https://doi.org/10.1242/jcs.062414.

Al-Bassam J, Kim H, Brouhard G, van Oijen A, Harrison SC, Chang F. CLASP promotes microtubule rescue by recruiting tubulin dimers to the microtubule. Dev Cell 2010;19:245–58. https://doi.org/10.1016/j.devcel.2010.07.016.

Al-Bassam J, Larsen NA, Hyman AA, Harrison SC. Crystal structure of a TOG domain: conserved features of XMAP215/Dis1-family TOG domains and implications for tubulin binding. Struct Lond Engl 1993 2007;15:355–62. https://doi.org/10.1016/j.str.2007.01.012.

Al-Bassam J, Ozer RS, Safer D, Halpain S, Milligan RA. MAP2 and tau bind longitudinally along the outer ridges of microtubule protofilaments. J Cell Biol 2002;157:1187–96. https://doi.org/10.1083/jcb.200201048.

Alblazi KMO, Siar CH. Cellular protrusions—lamellipodia, filopodia, invadopodia and podosomes—and their roles in progression of orofacial tumours: current understanding. Asian Pac J Cancer Prev APJCP 2015;16:2187–91. https://doi.org/10.7314/apjcp.2015.16.6.2187.

Alderson NL, Wang Y, Blatnik M, Frizzell N, Walla MD, Lyons TJ, et al. S-(2-Succinyl)cysteine: a novel chemical modification of tissue proteins by a Krebs cycle intermediate. Arch Biochem Biophys 2006;450:1–8. https://doi.org/10.1016/j.abb.2006.03.005.

Alli E, Bash-Babula J, Yang J-M, Hait WN. Effect of stathmin on the sensitivity to antimicrotubule drugs in human breast cancer. Cancer Res 2002;62:6864–9.

Altieri DC. Survivin, versatile modulation of cell division and apoptosis in cancer. Oncogene 2003;22:8581–9. https://doi.org/10.1038/sj.onc.1207113.

Amayed P, Pantaloni D, Carlier M-F. The effect of stathmin phosphorylation on microtubule assembly depends on tubulin critical concentration. J Biol Chem 2002;277:22718–24. https://doi.org/10.1074/jbc.M111605200.

Arnold J, Schattschneider J, Blechner C, Krisp C, Schlüter H, Schweizer M, et al. Tubulin Tyrosine Ligase Like 4 (TTLL4) overexpression in breast cancer cells is associated with brain metastasis and alters exosome biogenesis. J Exp Clin Cancer Res CR 2020;39:205. https://doi.org/10.1186/s13046-020-01712-w.

Artym VV, Zhang Y, Seillier-Moiseiwitsch F, Yamada KM, Mueller SC. Dynamic interactions of cortactin and membrane type 1 matrix metalloproteinase at invadopodia: defining the stages of invadopodia formation and function. Cancer Res 2006;66:3034–43. https://doi.org/10.1158/0008-5472.CAN-05-2177.

Avila J. Microtubule functions. Life Sci 1992;50:327–34. https://doi.org/10.1016/0024-3205(92)90433-p.

Banerjee A. Increased levels of tyrosinated alpha-, beta(III)-, and beta(IV)-tubulin isotypes in paclitaxel-resistant MCF-7 breast cancer cells. Biochem Biophys Res Commun 2002;293:598–601. https://doi.org/10.1016/S0006-291X(02)00269-3.

Barbier P, Zejneli O, Martinho M, Lasorsa A, Belle V, Smet-Nocca C, et al. Role of Tau as a Microtubule-Associated Protein: Structural and Functional Aspects. Front Aging Neurosci 2019;11:204. https://doi.org/10.3389/fnagi.2019.00204.

Barbolina MV. Targeting Microtubule-Associated Protein Tau in Chemotherapy-Resistant Models of High-Grade Serous Ovarian Carcinoma. Cancers 2022;14:4535. https://doi.org/10.3390/cancers14184535.

Barra HS, Arcce CA, Rodriguez JA, Caputto R. Uncorporation of phenylalanine as a single unit into rat brain protein: reciprocal inhibition by phenylalanine and tyrosine of their respective incorporations. J Neurochem 1973a;21:1241–51. https://doi.org/10.1111/j.1471-4159.1973.tb07578.x.

Barra HS, Arce CA, Rodríguez JA, Caputto R. Some common properties of the protein that incorporates tyrosine as a single unit and the microtubule proteins. Biochem Biophys Res Commun 1974;60:1384–90. https://doi.org/10.1016/0006-291x(74)90351-9.

Barra HS, Rodriguez JA, Arce CA, Caputto R. A soluble preparation from rat brain that incorporates into its own proteins ( 14 C)arginine by a ribonuclease-sensitive system and ( 14 C)tyrosine by a ribonuclease-insensitive system. J Neurochem 1973b;20:97–108. https://doi.org/10.1111/j.1471-4159.1973.tb12108.x.

Berges R, Baeza-Kallee N, Tabouret E, Chinot O, Petit M, Kruczynski A, et al. End-binding 1 protein overexpression correlates with glioblastoma progression and sensitizes to Vinca-alkaloids in vitro and in vivo. Oncotarget 2014;5:12769–87. https://doi.org/10.18632/oncotarget.2646.

Bieling P, Laan L, Schek H, Munteanu EL, Sandblad L, Dogterom M, et al. Reconstitution of a microtubule plus-end tracking system in vitro. Nature 2007;450:1100–5. https://doi.org/10.1038/nature06386.

Bobinnec Y, Moudjou M, Fouquet JP, Desbruyères E, Eddé B, Bornens M. Glutamylation of centriole and cytoplasmic tubulin in proliferating non-neuronal cells. Cell Motil Cytoskeleton 1998;39:223–32. https://doi.org/10.1002/(SICI)1097-0169(1998)39:3<223::AID--CM5>3.0.CO;2-5.

Bosch Grau M, Gonzalez Curto G, Rocha C, Magiera MM, Marques Sousa P, Giordano T, et al. Tubulin glycylases and glutamylases have distinct functions in stabilization and motility of ependymal cilia. J Cell Biol 2013;202:441–51. https://doi.org/10.1083/jcb.201305041.

Brady ST, Tytell M, Lasek RJ. Axonal tubulin and axonal microtubules: biochemical evidence for cold stability. J Cell Biol 1984;99:1716–24. https://doi.org/10.1083/jcb.99.5.1716.

Brouhard GJ, Stear JH, Noetzel TL, Al-Bassam J, Kinoshita K, Harrison SC, et al. XMAP215 is a processive microtubule polymerase. Cell 2008;132:79–88. https://doi.org/10.1016/j.cell.2007.11.043.

Bulinski JC, Borisy GG. Widespread distribution of a 210,000 mol wt microtubule-associated protein in cells and tissues of primates. J Cell Biol 1980;87:802–8. https://doi.org/10.1083/jcb.87.3.802.

Bulinski JC, McGraw TE, Gruber D, Nguyen HL, Sheetz MP. Overexpression of MAP4 inhibits organelle motility and trafficking in vivo. J Cell Sci 1997;110 (Pt 24):3055–64. https://doi.org/10.1242/jcs.110.24.3055.

Carvalho P, Gupta ML, Hoyt MA, Pellman D. Cell cycle control of kinesin-mediated transport of Bik1 (CLIP-170) regulates microtubule stability and dynein activation. Dev Cell 2004;6:815–29. https://doi.org/10.1016/j.devcel.2004.05.001.

Cassimeris L. The oncoprotein 18/stathmin family of microtubule destabilizers. Curr Opin Cell Biol 2002;14:18–24. https://doi.org/10.1016/s0955-0674(01)00289-7.

Cassimeris L, Morabito J. TOGp, the human homolog of XMAP215/Dis1, is required for centrosome integrity, spindle pole organization, and bipolar spindle assembly. Mol Biol Cell 2004;15:1580–90. https://doi.org/10.1091/mbc.e03-07-0544.

Cassimeris L, Spittle C. Regulation of microtubule-associated proteins. Int Rev Cytol 2001;210:163–226. https://doi.org/10.1016/s0074-7696(01)10006-9.

Cheung CHA, Chen H-H, Kuo C-C, Chang C-Y, Coumar MS, Hsieh H-P, et al. Survivin counteracts the therapeutic effect of microtubule de-stabilizers by stabilizing tubulin polymers. Mol Cancer 2009;8:43. https://doi.org/10.1186/1476-4598-8-43.

Chu C-W, Hou F, Zhang J, Phu L, Loktev AV, Kirkpatrick DS, et al. A novel acetylation of β-tubulin by San modulates microtubule polymerization via down-regulating tubulin incorporation. Mol Biol Cell 2011;22:448–56. https://doi.org/10.1091/mbc.E10-03-0203.

Chung WJ, Kindler S, Seidenbecher C, Garner CC. MAP2a, an alternatively spliced variant of microtubule-associated protein 2. J Neurochem 1996;66:1273–81. https://doi.org/10.1046/j.1471-4159.1996.66031273.x.

Cohen C, Lohmann CM, Cotsonis G, Lawson D, Santoianni R. Survivin expression in ovarian carcinoma: correlation with apoptotic markers and prognosis. Mod Pathol Off J U S Can Acad Pathol Inc 2003;16:574–83. https://doi.org/10.1097/01.MP.0000073868.31297.B0.

Davenport AM, Collins LN, Chiu H, Minor PJ, Sternberg PW, Hoelz A. Structural and functional characterization of the α-tubulin acetyltransferase MEC-17. J Mol Biol 2014;426:2605–16. https://doi.org/10.1016/j.jmb.2014.05.009.

David-Pfeuty T, Singer SJ. Altered distributions of the cytoskeletal proteins vinculin and alpha-actinin in cultured fibroblasts transformed by Rous sarcoma virus. Proc Natl Acad Sci U S A 1980;77:6687–91. https://doi.org/10.1073/pnas.77.11.6687.

Dehmelt L, Halpain S. The MAP2/Tau family of microtubule-associated proteins. Genome Biol 2005;6:204. https://doi.org/10.1186/gb-2004-6-1-204.

Dehmelt L, Halpain S. Actin and microtubules in neurite initiation: are MAPs the missing link? J Neurobiol 2004;58:18–33. https://doi.org/10.1002/neu.10284.

Domnitz SB, Wagenbach M, Decarreau J, Wordeman L. MCAK activity at microtubule tips regulates spindle microtubule length to promote robust kinetochore attachment. J Cell Biol 2012;197:231–7. https://doi.org/10.1083/jcb.201108147.

Don S, Verrills NM, Liaw TYE, Liu MLM, Norris MD, Haber M, et al. Neuronal-associated microtubule proteins class III beta-tubulin and MAP2c in neuroblastoma: role in resistance to microtubule-targeted drugs. Mol Cancer Ther 2004;3:1137–46.

Dong X, Liu F, Sun L, Liu M, Li D, Su D, et al. Oncogenic function of microtubule end-binding protein 1 in breast cancer. J Pathol 2010;220:361–9. https://doi.org/10.1002/path.2662.

Downing KH, Nogales E. Crystallographic structure of tubulin: implications for dynamics and drug binding. Cell Struct Funct 1999;24:269–75. https://doi.org/10.1247/csf.24.269.

Dye RB, Fink SP, Williams RC. Taxol-induced flexibility of microtubules and its reversal by MAP-2 and Tau. J Biol Chem 1993;268:6847–50.

Eli S, Castagna R, Mapelli M, Parisini E. Recent Approaches to the Identification of Novel Microtubule-Targeting Agents. Front Mol Biosci 2022;9:841777. https://doi.org/10.3389/fmolb.2022.841777.

Eshun-Wilson, L., Zhang, R., Portran, D. Effects of alpha-tubulin acetylation on microtubule structure and stability. Proc Natl Acad Sci 2019;116:10366–71.

Fang D, Hallman J, Sangha N, Kute TE, Hammarback JA, White WL, et al. Expression of microtubule-associated protein 2 in benign and malignant melanocytes: implications for differentiation and progression of cutaneous melanoma. Am J Pathol 2001;158:2107–15. https://doi.org/10.1016/S0002-9440(10)64682-2.

Faruki S, Geahlen RL, Asai DJ. Syk-dependent phosphorylation of microtubules in activated B-lymphocytes. J Cell Sci 2000; 113 (Pt 14):2557–65. https://doi.org/10.1242/jcs.113.14.2557.

Felgner H, Frank R, Biernat J, Mandelkow EM, Mandelkow E, Ludin B, et al. Domains of neuronal microtubule-associated proteins and flexural rigidity of microtubules. J Cell Biol 1997;138:1067–75. https://doi.org/10.1083/jcb.138.5.1067.

Fernandez JA, Keshvara LM, Peters JD, Furlong MT, Harrison ML, Geahlen RL. Phosphorylation- and activation-independent association of the tyrosine kinase Syk and the tyrosine kinase substrates Cbl and Vav with tubulin in B-cells. J Biol Chem 1999;274:1401–6. https://doi.org/10.1074/jbc.274.3.1401.

Ferreira LT, Figueiredo AC, Orr B, Lopes D, Maiato H. Dissecting the role of the tubulin code in mitosis. Methods Cell Biol 2018;144:33–74. https://doi.org/10.1016/bs.mcb.2018.03.040.

Fourest-Lieuvin A, Peris L, Gache V, Garcia-Saez I, Juillan-Binard C, Lantez V, et al. Microtubule regulation in mitosis: tubulin phosphorylation by the cyclin-dependent kinase Cdk1. Mol Biol Cell 2006;17:1041–50. https://doi.org/10.1091/mbc.e05-07-0621.

Gadadhar S, Alvarez Viar G, Hansen JN, Gong A, Kostarev A, Ialy-Radio C, et al. Tubulin glycylation controls axonemal dynein activity, flagellar beat, and male fertility. Science 2021;371. https://doi.org/10.1126/science.abd4914.

Gadadhar S, Dadi H, Bodakuntla S, Schnitzler A, Bièche I, Rusconi F, et al. Tubulin glycylation controls primary cilia length. J Cell Biol 2017;216:2701–13. https://doi.org/10.1083/jcb.201612050.

Galjart N. CLIPs and CLASPs and cellular dynamics. Nat Rev Mol Cell Biol 2005;6:487–98. https://doi.org/10.1038/nrm1664.

Gamblin TC, Nachmanoff K, Halpain S, Williams RC. Recombinant microtubule-associated protein 2c reduces the dynamic instability of individual microtubules. Biochemistry 1996;35:12576–86. https://doi.org/10.1021/bi961135d.

Gard DL, Kirschner MW. A polymer-dependent increase in phosphorylation of beta-tubulin accompanies differentiation of a mouse neuroblastoma cell line. J Cell Biol 1985;100:764–74. https://doi.org/10.1083/jcb.100.3.764.

Garner CC, Brugg B, Matus A. A 70-kilodalton microtubule-associated protein (MAP2c), related to MAP2. J Neurochem 1988;50:609–15. https://doi.org/10.1111/j.1471-4159.1988.tb02954.x.

Génot E, Gligorijevic B. Invadosomes in their natural habitat. Eur J Cell Biol 2014;93:367–79. https://doi.org/10.1016/j.ejcb.2014.10.002.

Goodman DB, Rasmussen H, DiBella F, Guthrow CE. Cyclic adenosine 3':5'-monophosphate-stimulated phosphorylation of isolated neurotubule subunits. Proc Natl Acad Sci U S A 1970;67:652–9. https://doi.org/10.1073/pnas.67.2.652.

Goodson HV, Jonasson EM. Microtubules and Microtubule-Associated Proteins. Cold Spring Harb Perspect Biol 2018;10:a022608. https://doi.org/10.1101/cshperspect.a022608.

Gupta KK, Li C, Duan A, Alberico EO, Kim OV, Alber MS, et al. Mechanism for the catastrophe-promoting activity of the microtubule destabilizer Op18/stathmin. Proc Natl Acad Sci U S A 2013;110:20449–54. https://doi.org/10.1073/pnas.1309958110.

Gurler H, Yu Y, Choi J, Kajdacsy-Balla AA, Barbolina MV. Three-dimensional collagen type I matrix up-regulates nuclear isoforms of the microtubule associated protein tau implicated in resistance to paclitaxel therapy in ovarian carcinoma. Int J Mol Sci 2015;16:3419–33. https://doi.org/10.3390/ijms16023419.

Hammond JW, Cai D, Verhey KJ. Tubulin modifications and their cellular functions. Curr Opin Cell Biol 2008;20:71–6. https://doi.org/10.1016/j.ceb.2007.11.010.

Hayashi I, Ikura M. Crystal structure of the amino-terminal microtubule-binding domain of end-binding protein 1 (EB1). J Biol Chem 2003;278:36430–4. https://doi.org/10.1074/jbc.M305773200.

Honnappa S, Gouveia SM, Weisbrich A, Damberger FF, Bhavesh NS, Jawhari H, et al. An EB1-binding motif acts as a microtubule tip localization signal. Cell 2009;138:366–76. https://doi.org/10.1016/j.cell.2009.04.065.

Honnappa S, Okhrimenko O, Jaussi R, Jawhari H, Jelesarov I, Winkler FK, et al. Key interaction modes of dynamic +TIP networks. Mol Cell 2006;23:663–71. https://doi.org/10.1016/j.molcel.2006.07.013.

Ichihara K, Kitazawa H, Iguchi Y, Hotani H, Itoh TJ. Visualization of the stop of microtubule depolymerization that occurs at the high-density region of microtubule-associated protein 2 (MAP2). J Mol Biol 2001;312:107–18. https://doi.org/10.1006/jmbi.2001.4934.

Illenberger S, Drewes G, Trinczek B, Biernat J, Meyer HE, Olmsted JB, et al. Phosphorylation of microtubule-associated proteins MAP2 and MAP4 by the protein kinase p110mark. Phosphorylation sites and regulation of microtubule dynamics. J Biol Chem 1996;271:10834–43. https://doi.org/10.1074/jbc.271.18.10834.

Janke C, Rogowski K, van Dijk J. Polyglutamylation: a fine-regulator of protein function? "Protein Modifications: beyond the usual suspects" review

series. EMBO Rep 2008;9:636–41. https://doi.org/10.1038/embor.2008.114.
Janke C, Rogowski K, Wloga D, Regnard C, Kajava AV, Strub J-M, et al. Tubulin polyglutamylase enzymes are members of the TTL domain protein family. Science 2005;308:1758–62. https://doi.org/10.1126/science.1113010.
Ji S, Kang JG, Park SY, Lee J, Oh YJ, Cho JW. O-GlcNAcylation of tubulin inhibits its polymerization. Amino Acids 2011;40:809–18. https://doi.org/10.1007/s00726-010-0698-9.
Kadowaki T, Fujita-Yamaguchi Y, Nishida E, Takaku F, Akiyama T, Kathuria S, et al. Phosphorylation of tubulin and microtubule-associated proteins by the purified insulin receptor kinase. J Biol Chem 1985;260:4016–20.
Kann M-L, Soues S, Levilliers N, Fouquet J-P. Glutamylated tubulin: diversity of expression and distribution of isoforms. Cell Motil Cytoskeleton 2003;55:14–25. https://doi.org/10.1002/cm.10107.
Kavallaris M. Microtubules and resistance to tubulin-binding agents. Nat Rev Cancer 2010;10:194–204. https://doi.org/10.1038/nrc2803.
Kavallaris M, Tait AS, Walsh BJ, He L, Horwitz SB, Norris MD, et al. Multiple microtubule alterations are associated with Vinca alkaloid resistance in human leukemia cells. Cancer Res 2001;61:5803–9.
Khoury H, Dankort DL, Sadekova S, Naujokas MA, Muller WJ, Park M. Distinct tyrosine autophosphorylation sites mediate induction of epithelial mesenchymal like transition by an activated ErbB-2/Neu receptor. Oncogene 2001;20:788–99. https://doi.org/10.1038/sj.onc.1204166.
Komarova Y, De Groot CO, Grigoriev I, Gouveia SM, Munteanu EL, Schober JM, et al. Mammalian end binding proteins control persistent microtubule growth. J Cell Biol 2009;184:691–706. https://doi.org/10.1083/jcb.200807179.
Kumar P, Wittmann T. +TIPs: SxIPping along microtubule ends. Trends Cell Biol 2012;22:418–28. https://doi.org/10.1016/j.tcb.2012.05.005.
Lansbergen G, Grigoriev I, Mimori-Kiyosue Y, Ohtsuka T, Higa S, Kitajima I, et al. CLASPs attach microtubule plus ends to the cell cortex through a complex with LL5beta. Dev Cell 2006;11:21–32. https://doi.org/10.1016/j.devcel.2006.05.012.
Lauffart B, Vaughan MM, Eddy R, Chervinsky D, DiCioccio RA, Black JD, et al. Aberrations of TACC1 and TACC3 are associated with ovarian cancer. BMC Womens Health 2005;5:8. https://doi.org/10.1186/1472-6874-5-8.
Liao S, Rajendraprasad G, Wang N, Eibes S, Gao J, Yu H, et al. Molecular basis of vasohibins-mediated detyrosination and its impact on spindle function and mitosis. Cell Res 2019;29:533–47. https://doi.org/10.1038/s41422-019-0187-y.
Ligon LA, Shelly SS, Tokito M, Holzbaur ELF. The microtubule plus-end proteins EB1 and dynactin have differential effects on microtubule polymerization. Mol Biol Cell 2003;14:1405–17. https://doi.org/10.1091/mbc.e02-03-0155.
Liu Y, Sturgis CD, Grzybicki DM, Jasnosz KM, Olson PR, Tong M, et al. Microtubule-associated protein-2: a new sensitive and specific marker for pulmonary carcinoid tumor and small cell carcinoma. Mod Pathol Off J U S Can Acad Pathol Inc 2001;14:880–5. https://doi.org/10.1038/modpathol.3880406.
Liu F, Sun Y-L, Xu Y, Liu Fang, Wang L-S, Zhao X-H. Expression and phosphorylation of stathmin correlate with cell migration in esophageal squamous cell carcinoma. Oncol Rep 2013;29:419–24. https://doi.org/10.3892/or.2012.2157.
Luo Y, Li D, Ran J, Yan B, Chen J, Dong X, et al. End-binding protein 1 stimulates paclitaxel sensitivity in breast cancer by promoting its actions toward microtubule assembly and stability. Protein Cell 2014;5:469–79. https://doi.org/10.1007/s13238-014-0053-0.
Ma X, Sayeski PP. Identification of tubulin as a substrate of Jak2 tyrosine kinase and its role in Jak2-dependent signaling. Biochemistry 2007;46:7153–62. https://doi.org/10.1021/bi700101n.
Maiato H, Sampaio P, Sunkel CE. Microtubule-associated proteins and their essential roles during mitosis. Int Rev Cytol 2004;241:53–153. https://doi.org/10.1016/S0074-7696(04)41002-X.
Markovsky E, de Stanchina E, Itzkowitz A, Haimovitz-Friedman A, Rotenberg SA. Phosphorylation State of Ser165 in α-Tubulin is a Toggle Switch That Controls Proliferating Human Breast Tumors. Cell Signal 2018;52:74–82. https://doi.org/10.1016/j.cellsig.2018.08.021.
Matov A, Applegate K, Kumar P, Thoma C, Krek W, Danuser G, et al. Analysis of microtubule dynamic instability using a plus-end growth marker. Nat Methods 2010;7:761–8. https://doi.org/10.1038/nmeth.1493.
Miller LM, Menthena A, Chatterjee C, Verdier-Pinard P, Novikoff PM, Horwitz SB, et al. Increased levels of a unique post-translationally modified betaIVb-tubulin isotype in liver cancer. Biochemistry 2008;47:7572–82. https://doi.org/10.1021/bi8005225.
Mimori-Kiyosue Y, Shiina N, Tsukita S. The dynamic behavior of the APC-binding protein EB1 on the distal ends of microtubules. Curr Biol CB 2000;10:865–8. https://doi.org/10.1016/s0960-9822(00)00600-x.
Murphy M, Hinman A, Levine AJ. Wild-type p53 negatively regulates the expression of a microtubule-associated protein. Genes Dev 1996;10:2971–80. https://doi.org/10.1101/gad.10.23.2971.
Noiges R, Eichinger R, Kutschera W, Fischer I, Németh Z, Wiche G, Propst F. Microtubule-Associated Protein 1A (MAP1A) and MAP1B: Light Chains Determine Distinct Functional Properties. J Neurosci 2002;22(6):2106–14. https://doi.org/10.1523/JNEUROSCI.22-06-02106.2002.
Ozer RS, Halpain S. Phosphorylation-dependent localization of microtubule-associated protein MAP2c to the actin cytoskeleton. Mol Biol Cell 2000;11:3573–87. https://doi.org/10.1091/mbc.11.10.3573.
Palazzo A, Ackerman B, Gundersen GG. Cell biology: Tubulin acetylation and cell motility. Nature 2003;421:230. https://doi.org/10.1038/421230a.

Parysek LM, Asnes CF, Olmsted JB. MAP 4: occurrence in mouse tissues. J Cell Biol 1984a;99:1309–15. https://doi.org/10.1083/jcb.99.4.1309.

Parysek LM, Wolosewick JJ, Olmsted JB. MAP 4: a microtubule-associated protein specific for a subset of tissue microtubules. J Cell Biol 1984b;99:2287–96. https://doi.org/10.1083/jcb.99.6.2287.

Penet M-F, Krishnamachary B, Wildes FB, Mironchik Y, Hung C-F, Wu T, et al. Ascites Volumes and the Ovarian Cancer Microenvironment. Front Oncol 2018;8:595. https://doi.org/10.3389/fonc.2018.00595.

Perez F, Diamantopoulos GS, Stalder R, Kreis TE. CLIP-170 highlights growing microtubule ends in vivo. Cell 1999;96:517–27. https://doi.org/10.1016/s0092-8674(00)80656-x.

Piehl M, Cassimeris L. Organization and dynamics of growing microtubule plus ends during early mitosis. Mol Biol Cell 2003;14:916–25. https://doi.org/10.1091/mbc.e02-09-0607.

Pierre P, Scheel J, Rickard JE, Kreis TE. CLIP-170 links endocytic vesicles to microtubules. Cell 1992;70:887–900. https://doi.org/10.1016/0092-8674(92)90240-d.

Poruchynsky MS, Giannakakou P, Ward Y, Bulinski JC, Telford WG, Robey RW, et al. Accompanying protein alterations in malignant cells with a microtubule-polymerizing drug-resistance phenotype and a primary resistance mechanism. Biochem Pharmacol 2001;62:1469–80. https://doi.org/10.1016/s0006-2952(01)00804-8.

Raj D, Liu T, Samadashwily G, Li F, Grossman D. Survivin repression by p53, Rb and E2F2 in normal human melanocytes. Carcinogenesis 2008;29:194–201. https://doi.org/10.1093/carcin/bgm219.

Raybin D, Flavin M. Enzyme which specifically adds tyrosine to the alpha chain of tubulin. Biochemistry 1977;16:2189–94. https://doi.org/10.1021/bi00629a023.

Reader J. Microtentacle formation in ovarian carcinoma. Cancers 2021;(in revision for publication).

Redeker V, Levilliers N, Schmitter JM, Le Caer JP, Rossier J, Adoutte A, et al. Polyglycylation of tubulin: a post-translational modification in axonemal microtubules. Science 1994;266:1688–91. https://doi.org/10.1126/science.7992051.

Rickard JE, Kreis TE. Identification of a novel nucleotide-sensitive microtubule-binding protein in HeLa cells. J Cell Biol 1990;110:1623–33. https://doi.org/10.1083/jcb.110.5.1623.

Robert M, Mathuranath PS. Tau and tauopathies. Neurol India 2007;55:11–6. https://doi.org/10.4103/0028-3886.30420.

Rocha C, Papon L, Cacheux W, Marques Sousa P, Lascano V, Tort O, et al. Tubulin glycylases are required for primary cilia, control of cell proliferation and tumor development in colon. EMBO J 2014;33:2247–60. https://doi.org/10.15252/embj.201488466.

Rogowski K, van Dijk J, Magiera MM, Bosc C, Deloulme J-C, Bosson A, et al. A family of protein-deglutamylating enzymes associated with neurodegeneration. Cell 2010;143:564–78. https://doi.org/10.1016/j.cell.2010.10.014.

Roll-Mecak A, McNally FJ. Microtubule-severing enzymes. Curr Opin Cell Biol 2010;22:96–103. https://doi.org/10.1016/j.ceb.2009.11.001.

Rosa J, Canovas P, Islam A, Altieri DC, Doxsey SJ. Survivin modulates microtubule dynamics and nucleation throughout the cell cycle. Mol Biol Cell 2006;17:1483–93. https://doi.org/10.1091/mbc.e05-08-0723.

Rubin CI, Atweh GF. The role of stathmin in the regulation of the cell cycle. J Cell Biochem 2004;93:242–50. https://doi.org/10.1002/jcb.20187.

Schiewek J, Schumacher U, Lange T, Joosse SA, Wikman H, Pantel K, et al. Clinical relevance of cytoskeleton associated proteins for ovarian cancer. J Cancer Res Clin Oncol 2018;144:2195–205. https://doi.org/10.1007/s00432-018-2710-9.

Schuyler SC, Pellman D. Microtubule "plus-end-tracking proteins": The end is just the beginning. Cell 2001;105:421–4. https://doi.org/10.1016/s0092-8674(01)00364-6.

Smoter M, Bodnar L, Grala B, Stec R, Zieniuk K, Kozlowski W, et al. Tau protein as a potential predictive marker in epithelial ovarian cancer patients treated with paclitaxel/platinum first-line chemotherapy. J Exp Clin Cancer Res CR 2013;32:25. https://doi.org/10.1186/1756-9966-32-25.

Steffensen KD, Smoter M, Waldstrøm M, Grala B, Bodnar L, Stec R, et al. Resistance to first line platinum paclitaxel chemotherapy in serous epithelial ovarian cancer: the prediction value of ERCC1 and Tau expression. Int J Oncol 2014;44:1736–44. https://doi.org/10.3892/ijo.2014.2311.

Su D, Smith SM, Preti M, Schwartz P, Rutherford TJ, Menato G, et al. Stathmin and tubulin expression and survival of ovarian cancer patients receiving platinum treatment with and without paclitaxel. Cancer 2009;115:2453–63. https://doi.org/10.1002/cncr.24282.

Sun X, Li D, Yang Y, Ren Y, Li J, Wang Z, et al. Microtubule-binding protein CLIP-170 is a mediator of paclitaxel sensitivity. J Pathol 2012;226:666–73. https://doi.org/10.1002/path.3026.

Suraneni P, Rubinstein B, Unruh JR, Durnin M, Hanein D, Li R. The Arp2/3 complex is required for lamellipodia extension and directional fibroblast cell migration. J Cell Biol 2012;197:239–51. https://doi.org/10.1083/jcb.201112113.

Szyk A, Deaconescu AM, Spector J, Goodman B, Valenstein ML, Ziolkowska NE, et al. Molecular basis for age-dependent microtubule acetylation by tubulin acetyltransferase. Cell 2014;157:1405–15. https://doi.org/10.1016/j.cell.2014.03.061.

Tanida I, Ueno T, Kominami E. LC3 and Autophagy. Methods Mol Biol Clifton NJ 2008;445:77–88. https://doi.org/10.1007/978-1-59745-157-4_4.

Thomas GE, Sreeja JS, Gireesh KK, Gupta H, Manna TK. +TIP EB1 downregulates paclitaxel-induced proliferation inhibition and apoptosis in breast cancer cells through inhibition of paclitaxel binding on microtubules. Int J Oncol 2015;46:133–46. https://doi.org/10.3892/ijo.2014.2701.

Torrino S, Grasset EM, Audebert S, Belhadj I, Lacoux C, Haynes M, et al. Mechano-induced cell metabolism promotes microtubule glutamylation to force metastasis. Cell Metab 2021;33:1342–1357.e10. https://doi.org/10.1016/j.cmet.2021.05.009.

Tournebize R, Popov A, Kinoshita K, Ashford AJ, Rybina S, Pozniakovsky A, et al. Control of microtubule dynamics by the antagonistic activities of XMAP215 and XKCM1 in Xenopus egg extracts. Nat Cell Biol 2000;2:13–9. https://doi.org/10.1038/71330.

Vandecandelaere A, Pedrotti B, Utton MA, Calvert RA, Bayley PM. Differences in the regulation of microtubule dynamics by microtubule-associated proteins MAP1B and MAP2. Cell Motil Cytoskeleton 1996;35:134–46. https://doi.org/10.1002/(SICI)1097--0169(1996)35:2<134::AID-CM6>3.0.CO;2-A.

Vasquez RJ, Gard DL, Cassimeris L. XMAP from Xenopus eggs promotes rapid plus end assembly of microtubules and rapid microtubule polymer turnover. J Cell Biol 1994;127:985–93. https://doi.org/10.1083/jcb.127.4.985.

Veitia R, David S, Barbier P, Vantard M, Gounon P, Bissery MC, et al. Proteolysis of microtubule associated protein 2 and sensitivity of pancreatic tumours to docetaxel. Br J Cancer 2000;83:544–9. https://doi.org/10.1054/bjoc.2000.1294.

Verrills NM, Flemming CL, Liu M, Ivery MT, Cobon GS, Norris MD, et al. Microtubule alterations and mutations induced by desoxyepothilone B: implications for drug-target interactions. Chem Biol 2003;10:597–607. https://doi.org/10.1016/s1074-5521(03)00141-8.

Wang Z-B, Peng X-Z, Chen S-S, Ning F-L, Du C-J, Wang K, et al. High p53 and MAP1 light chain 3A co-expression predicts poor prognosis in patients with esophageal squamous cell carcinoma. Mol Med Rep 2013;8:41–6. https://doi.org/10.3892/mmr.2013.1451.

Waters AM, Beales PL. Ciliopathies: an expanding disease spectrum. Pediatr Nephrol Berl Ger 2011;26:1039–56. https://doi.org/10.1007/s00467-010-1731-7.

Wattanathamsan O, Thararattanobon R, Rodsiri R, Chanvorachote P, Vinayanuwattikun C, Pongrakhananon V. Tubulin acetylation enhances lung cancer resistance to paclitaxel-induced cell death through Mcl-1 stabilization. Cell Death Discov 2021;7:67. https://doi.org/10.1038/s41420-021-00453-9.

Webster DR, Wehland J, Weber K, Borisy GG. Detyrosination of alpha tubulin does not stabilize microtubules in vivo. J Cell Biol 1990;111:113–22. https://doi.org/10.1083/jcb.111.1.113.

Weisbrich A, Honnappa S, Jaussi R, Okhrimenko O, Frey D, Jelesarov I, et al. Structure-function relationship of CAP-Gly domains. Nat Struct Mol Biol 2007;14:959–67. https://doi.org/10.1038/nsmb1291.

Whipple RA, Balzer EM, Cho EH, Matrone MA, Yoon JR, Martin SS. Vimentin filaments support extension of tubulin-based microtentacles in detached breast tumor cells. Cancer Res 2008;68:5678–88. https://doi.org/10.1158/0008-5472.CAN-07-6589.

Whipple RA, Cheung AM, Martin SS. Detyrosinated microtubule protrusions in suspended mammary epithelial cells promote reattachment. Exp Cell Res 2007;313:1326–36. https://doi.org/10.1016/j.yexcr.2007.02.001.

Wloga D, Joachimiak E, Fabczak H. Tubulin Post-Translational Modifications and Microtubule Dynamics. Int J Mol Sci 2017;18:2207. https://doi.org/10.3390/ijms18102207.

Xie S, Ogden A, Aneja R, Zhou J. Microtubule-Binding Proteins as Promising Biomarkers of Paclitaxel Sensitivity in Cancer Chemotherapy. Med Res Rev 2016;36:300–12. https://doi.org/10.1002/med.21378.

Xu Z, Schaedel L, Portran D, Aguilar A, Gaillard J, Marinkovich MP, et al. Microtubules acquire resistance from mechanical breakage through intralumenal acetylation. Science 2017;356:328–32. https://doi.org/10.1126/science.aai8764.

Yamauchi A, Kobayashi A, Oikiri H, Yokoyama Y. Functional role of the Tau protein in epithelial ovarian cancer cells. Reprod Med Biol 2017;16:143–51. https://doi.org/10.1002/rmb2.12019.

Yang S, Liang Y, Qian H, Li Q. TTLL12 expression in ovarian cancer correlates with a poor outcome. Int J Clin Exp Pathol 2020;13:239–47.

Yilmaz M, Christofori G. EMT, the cytoskeleton, and cancer cell invasion. Cancer Metastasis Rev 2009;28:15–33. https://doi.org/10.1007/s10555-008-9169-0.

Zhang CC, Yang JM, White E, Murphy M, Levine A, Hait WN. The role of MAP4 expression in the sensitivity to paclitaxel and resistance to vinca alkaloids in p53 mutant cells. Oncogene 1998;16:1617–24. https://doi.org/10.1038/sj.onc.1201658.

# The Impact of Centrosome Pathologies on Ovarian Cancer Development and Progression with a Focus on Centrosomes as Therapeutic Target

3

Heide Schatten

**Abstract**

The impact of centrosome abnormalities on cancer cell proliferation has been recognized as early as 1914 (Boveri, Zur Frage der Entstehung maligner Tumoren. Jena: G. Fisher, 1914), but vigorous research on molecular levels has only recently started when it became fully apparent that centrosomes can be targeted for new cancer therapies. While best known for their microtubule-organizing capabilities as MTOC (microtubule organizing center) in interphase and mitosis, centrosomes are now further well known for a variety of different functions, some of which are related to microtubule organization and consequential activities such as cell division, migration, maintenance of cell shape, and vesicle transport powered by motor proteins, while other functions include essential roles in cell cycle regulation, metabolic activities, signal transduction, proteolytic activity, and several others that are now heavily being investigated for their role in diseases and disorders (reviewed in Schatten and Sun, Histochem Cell Biol 150:303–325, 2018; Schatten, Adv Anat Embryol Cell Biol 235:43–50, 2022a; Schatten, Adv Anat Embryol Cell Biol 235:17–35, 2022b).

Cancer cell centrosomes differ from centrosomes in noncancer cells in displaying specific abnormalities that include phosphorylation abnormalities, overexpression of specific centrosomal proteins, abnormalities in centriole and centrosome duplication, formation of multipolar spindles that play a role in aneuploidy and genomic instability, and several others that are highlighted in the present review on ovarian cancer. Ovarian cancer cell centrosomes, like those in other cancers, display complex abnormalities that in part are based on the heterogeneity of cells in the cancer tissues resulting from different etiologies of individual cancer cells that will be discussed in more detail in this chapter.

Because of the critical role of centrosomes in cancer cell proliferation, several lines of research are being pursued to target centrosomes for therapeutic intervention to inhibit abnormal cancer cell proliferation and control tumor progression. Specific centrosome abnormalities observed in ovarian cancer will be addressed in this chapter with a focus on targeting such aberrations for ovarian cancer-specific therapies.

H. Schatten (✉)
University of Missouri-Columbia Department of Veterinary Pathobiology, Columbia, MO, USA
e-mail: SchattenH@missouri.edu



H. Schatten (ed.), *Cell and Molecular Biology of Ovarian Cancer*, Advances in Experimental Medicine and Biology 1452, https://doi.org/10.1007/978-3-031-58311-7_3

**Keywords**

Centrosomes · Centrosomal proteins · Centrosome phosphorylation · Centrosome clustering · Centrioles · Cell cycle · Microtubule organization · Multipolar spindles · Ovarian cancer development · Therapeutic targets · Cytoskeletal dysfunctions · Tumor suppressor genes · Primary cilia

## 3.1 Introduction

Ovarian cancer refers to cancers of the ovary, fallopian tubes, and peritoneum based on the origins of the specific tissue. Epithelial ovarian cancer is classified into five subtypes, primarily based on histology: high-grade serous carcinoma, endometrioid carcinoma, clear cell carcinoma, low-grade serous carcinoma, and mucinous carcinoma. Serous ovarian cancer is the most common type of epithelial ovarian cancer and accounts for about two-thirds of all epithelial ovarian cancers (Jayson et al. 2014).

Ovarian cancer is the fifth most common cause of death in US women and the leading cause of gynecologic cancer deaths. Data generated for 2020 show that there were about 21,750 new ovarian cancer cases, which comprises 1.2% of all cancer cases. Deaths related to ovarian cancer cases were 13,940 in 2020 (reviewed in Arora et al. 2023).

Multiple factors have been associated with an increased risk of ovarian cancer (reviewed in Arora et al. 2023), which includes inherited genetic mutations such as *BRCA1/BRCA2*. Inherited mutations are associated with about 5–15% of all ovarian carcinomas, while other mutations or abnormalities are due to different risk factors that can include lifestyle, dietary factors, exposure to toxic substances, and various others.

Early detection is still the best predictor of a positive outcome with 15% of cases being diagnosed at a local stage having a 92% 5-year survival rate while only a ca 60% 5-year survival rate is expected when the cancer is diagnosed with distal metastases which amounts to a 30.2% 5-year survival rate compared to 92.6% if detected at an early stage of spread.

Ovarian serous carcinoma is the most common type of ovarian carcinomas with about 10% presenting at low grade and 90% high grade. The low-grade subtype shows minimal nuclear atypia, rare mitosis, and lesser molecular abnormalities, while the high-grade subtype displays significant nuclear atypia and mitoses, with a significant high number of molecular abnormalities (Rosen et al. 2009; Arora et al. 2023).

The lack of potent therapies for ovarian cancer calls for new approaches to find new targets to combat the disease, a disease that is most commonly only detected at advanced stages (of initial diagnosis) when commonly available drugs are not effective in combatting the many aberrations associated with this highly heterogeneous disease. The 5-year survival rate of ca. 30% is significantly related to the high relapse rate. Therefore, new biomarkers for early detection and new therapies are needed to extend the survival time or eradicate the disease as an ultimate goal to combat ovarian cancer. More research is needed to find effective tumor markers and determine their functions in ovarian cancer in order to understand ovarian cancer pathogenesis and specific oncogenes and tumorigenesis for diagnosis, prevention, and treatment. Targeting cancer cell centrosomes offers a new approach to control cell proliferation and tumor spread.

While best known for their microtubule nucleating and organizing capabilities as MTOC (microtubule-organizing center) in interphase and mitosis, centrosomes are now further known for a variety of different functions, some of which are related to microtubule organization and consequential activities such as cell division, migration, maintenance of cell shape, and vesicle transport powered by motor proteins, while other functions include a significant role in cell cycle regulation, metabolic activities, signal transduction, proteolytic activity, and several others that are still being investigated for their role in diseases and disorders (reviewed in Schatten and Sun 2018; Dráber and Dráberová 2021; Schatten 2022a, b).

Based on in-depth centrosome research on cell, molecular, and genetic levels, we now know that in addition to the role as MTOC, centrosomes serve as important cellular communication hub for signal transduction pathways, holding key roles in cell cycle regulation and in several other complex cellular functions that directly or indirectly affect cell cycle progression and cellular metabolism. The centrosome further serves as docking station for enzyme-carrying vesicles that are translocated along microtubules, and it controls and directs translocation of macromolecular complexes and cellular organelles such as mitochondria (Figs. 3.1 and 3.2).

Centrosome abnormalities in cancer have been described as early as 1914 by the brilliant cytologist Theodore Boveri (1914) who implicated centrosome abnormalities in abnormal mitosis and impaired cell divisions that are hallmarks of cancer, thereby laying the intellectual and scientific groundwork for further explorations of the centrosome in cancer development and progression. Boveri's discoveries subsequently led to substantially more and significant discoveries especially during the past two decades. Boveri's groundbreaking work in 1914 (Boveri 1914) has been well recognized (reviewed in Schatten 2008; Schatten and Sun 2018). Furthermore, Boveri's microscope slides have been rediscovered and further analyzed (Scheer 2014, 2018), and because of the importance, his findings have been translated into English in recent years (Boveri 2008).

Numerous studies by now have shown that in normal mitosis, bipolar centrosomes form the bipolar mitotic spindle during a regular cell cycle, but in cancer cells, mitotic centrosome abnormalities of various kinds range from abnormal bipolar to multipolar spindles resulting in aneuploidy and genomic instability as consequences of such aberrations.

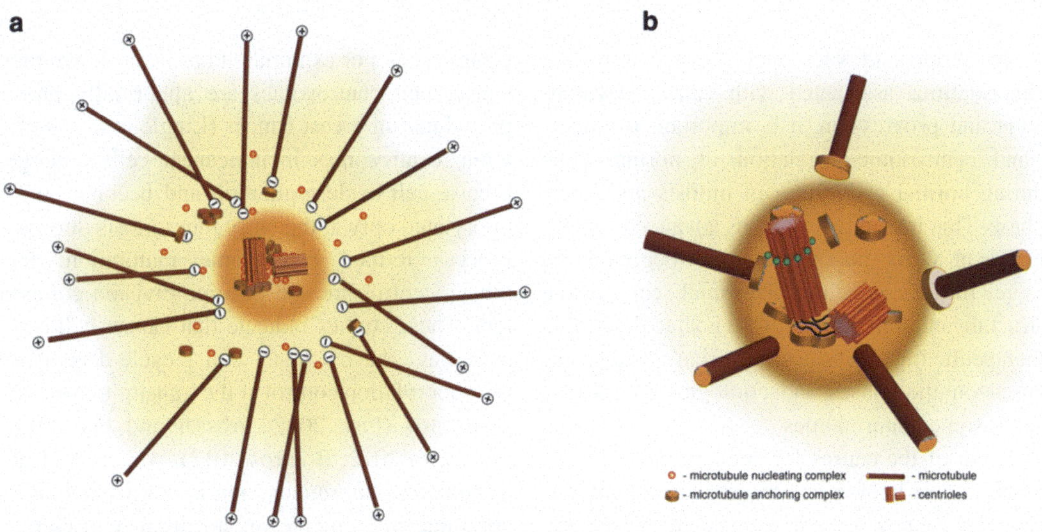


**Fig. 3.1** (**a**) A typical mammalian somatic cell centrosome consists of centrosomal material, also referred to as pericentriolar material (PCM), as it surrounds two perpendicularly oriented centrioles, termed mother and daughter centrioles. The mother centriole is distinguished from the daughter centriole by distal and subdistal appendages (shown as green necklace in (**b**). Both centrioles are connected by interconnected fibers (shown in **b**). The centrosomal material consists of a meshwork of proteins embedded in a matrix of yet undetermined structural composition. Gamma-tubulin and the gamma-tubulin ring complex are embedded in the centrosomal matrix and nucleate microtubules along with associated proteins with microtubule minus ends anchored to the matrix. This diagram also shows two complexes within the centrosomal matrix: the microtubule-nucleating complex and the microtubule anchoring complex. The diagram in **b** shows in more detail the two centrioles (mother and daughter centriole) in perpendicular orientation to each other. (Modified from Schatten (2013); reprinted from Schatten and Sun (2018))

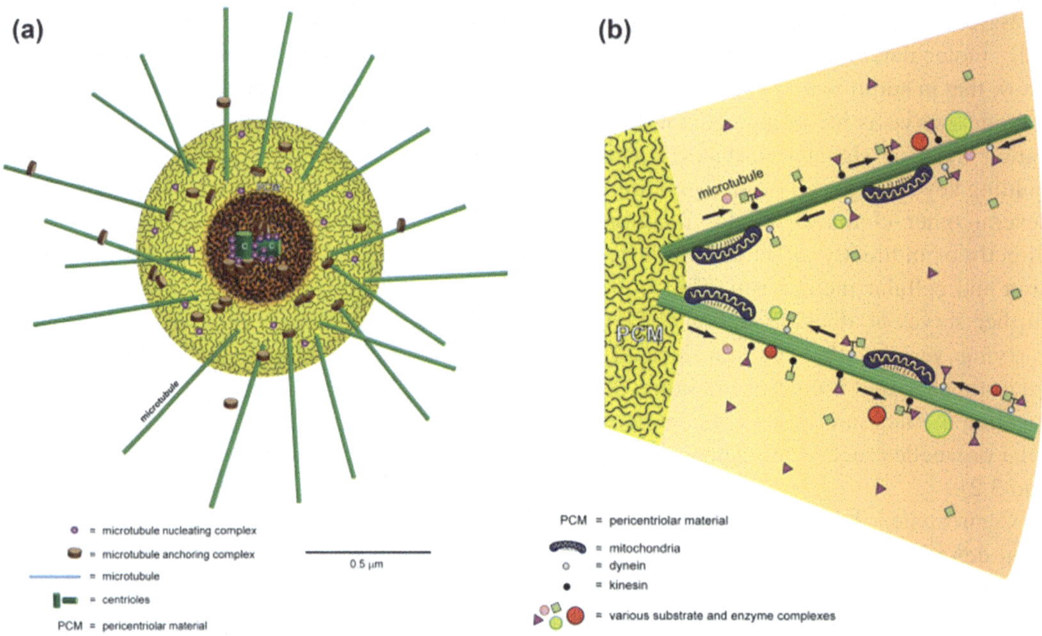


**Fig. 3.2** (**a**, **b**) Diagrammatic representation of a typical mammalian centrosome. The centrosome complex serves important functions as central station for signal transduction molecules (MAPK, Aurora A, Plk1, many other transient molecules) including cell organelles such as mitochondria

To understand such and other centrosome abnormalities associated with cancer development and progression, it is important to understand centrosome regulation in normal cells during normal cell cycles in mitosis and interphase. This line of research has advanced rapidly in recent years, and we are now beginning to more fully understand abnormal centrosome structure and functions with the goal to determine therapeutic interventions for ovarian cancer cures based on the causes and etiologies of specific centrosome abnormalities.

Some of the causes for centrosome dysfunctions are now known, and based on excellent new research, we now also know that centrosome dysfunctions are implicated in numerous diseases (reviewed in Badano et al. 2005; Schatten and Sun 2018). As centrosome abnormalities are severely associated with cancer development and progression (reviewed in Chan 2011), several labs are actively investigating cause and effect leading to centrosome abnormalities in cancer cells and to determine when centrosomes become dysfunctional during the cascade of events leading to the observed abnormalities (reviewed in Chan 2011). For example, it has been shown that cancer cell centrosomes are abnormally phosphorylated in breast cancer (Lingle et al. 1998), while centrosomes in noncancer cells undergo precise cell cycle regulation and become phosphorylated only at the entry into mitosis during a process termed centrosome maturation that allows centrosomes to become division competent. These studies indicate that cancer cell centrosomes have lost cell cycle-dependent phosphorylation control at the transition from G2 to mitosis (Fisk 2012; Prosser and Fry 2012; Fukasawa 2012; Boutros 2012), which can lead to numerous subsequent abnormalities including abnormal microtubule organization with consequences for aberrant segregation of chromosomes to the dividing daughter cells (Figs. 3.3 and 3.4) resulting in loss of tumor suppressor genes in some cells and increases in tumor promoter genes in others.

Other factors involved in cancer cell centrosome dysregulation include disruption of centrosome duplication (Korzeniewski and Duensing 2012), DNA damage caused by radiation (Saladino et al. 2012), protein degradation dys-

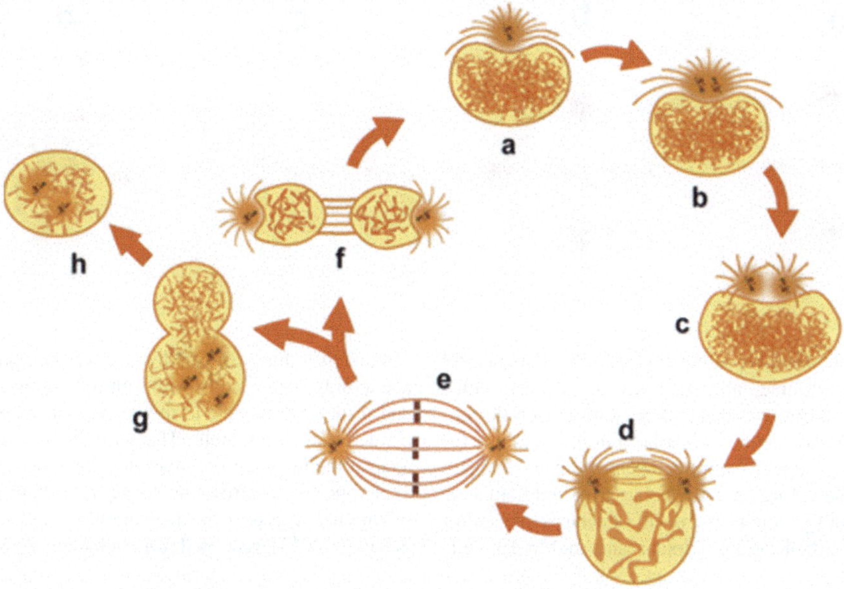


**Fig. 3.3** (**a–f**) Somatic cell centrosome cycle within the cell cycle. (**a**) The single interphase centrosome containing a pair of centrioles is closely associated with the nucleus and nucleates an array of interphase microtubules. (**b**) Centriole–centrosome duplication occurs during the S phase in synchrony with DNA duplication. (**c**) Separation of the duplicated centriole–centrosome complex toward the opposite spindle poles takes place in the early prophase stage. (**d**) The bipolar mitotic apparatus becomes established when each centriole–centrosome complex has reached the opposite pole, and the nuclear envelope has broken down. During this stage, interphase centrosomes mature into mitotic centrosomes acquiring mitosis-associated centrosomal proteins including NuMA that had moved out of the nucleus during nuclear envelope breakdown. (**e**) The metaphase centrosome becomes highly compacted to organize the metaphase spindle with microtubules attached to the kinetochores. (**f**) Telophase is the stage when centrosomal material becomes decompacted again before reorganizing into interphase centrosomes that associate with the nuclei of the separating daughter cells. (**g**, **h**) Centrosomal abnormalities associated with cell cycle dysfunctions. In cancer cells (**g**, **h**), centrosome and centriole numbers can amplify or over-replicate leading to aneuploidy or failure of cytokinesis. (Modified from Schatten (2013); reprinted from Schatten and Sun (2018))

functions (Fisk 2012; Prosser and Fry 2012; Kais and Parvin 2012), and numerous others including effects by environmental factors (Schatten and Sun 2009; Yan and Chng 2012; Olivero 2012; reviewed in Schatten 2008; Schatten and Sun 2018).

This chapter will focus on aberrant mechanisms leading to centrosome dysfunctions that are common to most tumor types but will highlight specific centrosome abnormalities observed in ovarian cancer that can present targets for ovarian cancer-specific therapies. This chapter will address (1) the structure, composition, and function of centrosomes and abnormalities/dysregulation in ovarian cancer; (2) centrosome regulation in normal cell cycles; (3) centrosome dysregulation, abnormalities in ovarian cancer, and new research to correct centrosome abnormalities; (4) the role of primary cilia in ovarian cancer; and (5) centrosomes as target for ovarian cancer therapy.

## 3.2 Structure, Composition, and Function of Centrosomes and Abnormalities/Dysregulation in Ovarian Cancer

The centrosome in a typical mammalian somatic cell consists of a perpendicularly oriented centriole pair that is embedded in a protein-rich centrosomal matrix (Figs. 3.1 and 3.2), also termed

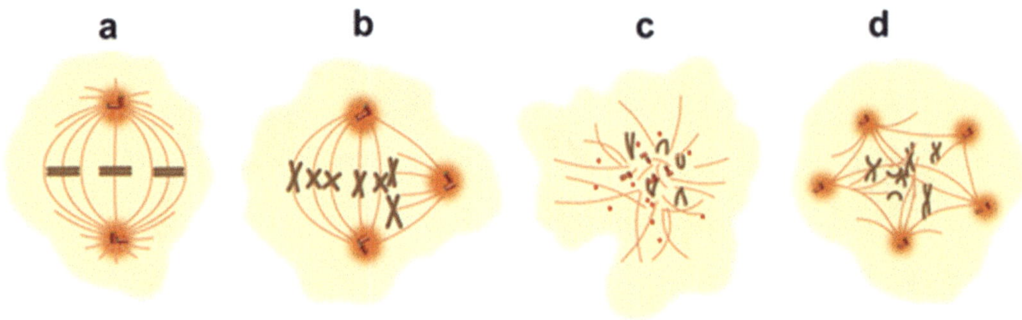


**Fig. 3.4** Schematic representation of various structural centriole–centrosome abnormalities in cancer cells. Cancer cell centrosomes can undergo various mitotic configurations forming either a bipolar mitotic apparatus (**a**) that oftentimes contains amplified centrosomes as a result of centrosome clustering and bipolar (**b**) or multipolar (**c**, **d**) mitotic configurations as a result of centrosome and/or centriole overduplication. Tripolar and multipolar cells can undergo abnormal cell divisions resulting in cells with aneuploidy, while cells with configurations shown in **c** and **d** may not be able to divide but rather become fragmented and disintegrate. The goal of therapies directed against the formation of abnormal centrosomes is to prevent abnormal centrosome clustering to induce fragmentation and cancer cell disintegration. (Modified from Schatten (2013); reprinted from Schatten and Sun (2018))

pericentriolar material (PCM). The centrosome organelle, unlike most other organelles, is not surrounded by a membrane that facilitates its multiple functions and structural and molecular remodeling within a typical cell cycle. The centrosomal matrix consists of a lattice of coiled-coil proteins, containing numerous specific centrosomal proteins including the γ-tubulin ring complexes (γ-TuRCs), pericentrin, and centrin (reviewed in Schatten 2008) that are among the best-studied centrosomal proteins so far. It also includes calcium-sensitive fibers (Salisbury 2004) and other structural elements that are still not yet well defined (reviewed in Schatten and Sun 2018). Scaffold and effector proteins, kinases, and phosphatases are associated with the PCM to form well-ordered layers that can be distinguished with super-resolution microscopy being involved in the organization of microtubules (Woodruff et al. 2014; Fry et al. 2017).

Cell cycle-specific centrosome dynamics includes remodeling of centrosomal proteins, which is especially pronounced during the transition from interphase to mitosis, the G2 to M transition, when γ-tubulin becomes increased to nucleate microtubules for the mitotic apparatus and when the nuclear mitotic apparatus protein (NuMA) becomes associated with the mitotic centrosome to form a crescent to aid in the organization of microtubules into the mitotic apparatus. Other cell cycle-specific remodeling of centrosome proteins includes posttranslation modifications that become important during cellular polarization (reviewed in Schatten and Sun 2014, 2018).

Centrioles, on the other hand, do not significantly change in their molecular composition throughout the cell cycle, but they are tightly synchronized with the centrosome cycle and undergo duplication in synchrony with the centrosomal matrix. For accurate composition of the centrosome complex, centrioles play important roles in the assembly of specific centrosome proteins and in the duplication of centrosomal material (Salisbury et al. 2002). Loss of synchronization is apparent in cancer cells, leading to abnormal cell divisions which will be discussed in more detail below in this chapter.

*Centrioles* in mammalian cells are composed of nine outer triplet microtubules forming a barrel-shaped small tube that does not contain central microtubules. Centriole duplication follows a semiconservative duplication pattern by which a younger (daughter) centriole forms perpendicular to the older (mother) centriole. The mother centriole is structurally and functionally distinguished from the daughter centriole in that the mother centriole contains appendages that become important for specific functions such as primary cilia formation that will be discussed in more detail below.

The cell cycle-specific regulation of the centriole–centrosome complex is critically important for its accurate functions, including the organization of accurate numbers of microtubule nucleation and organization that attach to kinetochores in mitotic cells and are part of a complex molecular machinery involved in the accurate separation of chromosomes to the daughter cells during cell division. Cell cycle abnormalities and their consequences leading to abnormal cell divisions as characteristic for cancer cells are shown in Fig. 3.3 (reviewed in Schatten and Sun 2014; Schatten 2022a, b).

Accurate coordination of the centrosome cycle with the chromosome cycle is critically important to precisely separate chromosomes during mitosis and cell division which includes accurate and synchronized duplication of the centrosome complex and chromosomes in parallel pathways of regulation to form the bipolar mitotic spindle that precisely partitions chromosomes equally to the daughter cells. In cancer cells, this coordination is lost, and centrosomes can form multiple poles (as shown in Fig. 3.4) that do not separate chromosomes equally to the dividing daughter cells but form cells with unequal chromosome numbers and cells that may lack tumor suppressor genes, thereby contributing to cancer cell growth and loss of tissue architecture. As a result, cellular structure and polarity become gradually lost with tumor progression and advanced histological grade.

In cancer cells, the γ-TuRC and associated proteins become dysregulated, resulting in abnormal mitoses that are hallmarks of cancer cells. Such abnormalities are being explored as targets for new cancer therapies to restore accurate molecular conditions, as will be discussed below.

As mentioned above, numerous centrosomal proteins are associated with the centrosome matrix that undergo cell cycle-specific regulation. Several hundred centrosomal proteins have been identified in specific cell cycle stages (Andersen et al. 2003), although a large number of these proteins are only temporarily associated with centrosomes and may be classified as centrosome-associated proteins. Others may use centrosomes as central hub for cell cycle-specific functions and include proteins of signal transduction cascades or the proteolytic pathway. This area of research is still actively being pursued, and so far, at least 60 centrosomal proteins have been identified in a typical somatic interphase cell centrosome (reviewed in Wilkinson et al. (2004)), although differences in composition and quantities may exist in different species (reviewed in Schatten (2022a, b)).

While centrosome core proteins are permanently associated with the centrosome structure, other proteins are part of the cell cycle-dependent centrosomal changes in most cell systems. Some of these centrosome proteins are overexpressed in cancer cells and play a role in centrosome amplification (discussed below).

Centrosome proteins that have been identified in purified centrosomes by mass spectrometric analysis include the *structural proteins* (alpha-tubulin, beta-tubulin, gamma-tubulin, gamma-tubulin complex components 1–6, centrin 2 and 3, AKAP450, pericentrin/kendrin, ninein, pericentriolar material 1 (PCM1), ch-TOG protein, C-Nap1, Cep250, Cep2, centriole-associated protein CEP110, Cep1, centriolin, centrosomal P4.1-associated protein (CPAP), CLIP-associating proteins CLASP1 and CLASP 2, ODF2, cenexin, Lis1, Nudel, EB1, centractin, myomegalin); *the regulatory molecules* (cell division protein 2 (Cdc2), Cdk1, cAMP-dependent protein kinase type II-alpha regulatory chain, cAMP-dependent protein kinase-alpha catalytic subunit, serine/threonine protein kinase Plk1, serine/threonine protein kinase Nek2, serine/threonine protein kinase Sak, Casein kinase I, delta and epsilon isoforms, protein phosphatase 2A, protein phosphatase 1 alpha isoform, 14-3-3 proteins, epsilon and gamma isoforms); the *motor and motor-related proteins* (dynein heavy chain, dynein intermediate chain, dynein light chain, dynactin 1, p150 Glued, dynactin 2, p50, dynactin 3); and the *heat shock proteins* (heat shock protein Hsp90, TCP subunits, and heat shock protein Hsp73).

The γ-tubulin ring complex, pericentrin, centrin, and the centrosome-associated protein NuMA (nuclear mitotic apparatus protein) have been shown to be aberrant in cancer cells, which is discussed in more detail as follows.

*Gamma-tubulin* is an essential centrosomal protein that is characteristically found in the centrosome core structure, but it can also serve as nucleating sites in areas other than the centrosome and includes the Golgi apparatus, the nuclear envelope, plasma membrane-associated sites, preexisting microtubules, and chromatin (reviewed in Schatten and Sun 2014; Schatten 2022a, b). The nucleating complex for microtubules from centrosomes is the ca. 2.2-MDa γ-TuRC that is associated with the centrosome core structure in all cells studied so far (Hannak et al. 2002) and consists of 12 or 14 γ-tubulin molecules. Several other components are needed to anchor the γ-TuRC to the centrosome core structure including the large coiled-coil A-kinase anchoring proteins (Tassin et al. 1998; Murphy et al. 1998; Dictenberg et al. 1998; Doxsey et al. 1994; Flory and Davis 2003; Gillingham and Munro 2000; Kawaguchi and Zheng 2004; Keryer et al. 2003; Steadman et al. 2002; Takahashi et al. 2002) and Cep135 (Ohta et al. 2002). The microtubule minus-end binding proteins including the γ-TuRC are also localized to the proximal ends of centrioles, while tubulin polyglutamylation of the centriole walls modulates interaction between tubulin- and microtubule-associated proteins (reviewed in Teixidó-Travesa et al. 2012; Schatten and Sun 2014).

Microtubule anchoring to the centrosome further includes ninein as a microtubule minus-end anchoring protein (Mogensen et al. 2000) and dynactin that plays a major role in microtubule anchorage at centrosomes as well as at non-centrosomal anchorage sites with preferential localization at the mother centriole where it also plays a role in microtubule organization (Quintyne et al. 1999; Schroer 2001; Quintyne and Schroer 2002).

In interphase, the γ-TuRC complex nucleates fewer but longer microtubules, while in mitosis, an increased number of γ-TuRCs becomes associated with the centrosome which is part of the centrosome maturation process that takes place from interphase to mitosis. Mitotic microtubules are shorter, larger in number, and highly dynamic to nucleate and organize the mitotic apparatus including microtubules attaching to the kinetochores (kinetochore microtubules). Mitotic centrosomes are regulated by a number of cell cycle-specific proteins, phosphatases and kinases, including the small GTPase Ran, Aurora kinase A, polo-like kinases.

Increased γ-tubulin expression has been reported for various cancers including laryngeal carcinomas (Syed et al. 2009), thyroid carcinomas (Montero-Conde et al. 2008), ovarian cancer (Hsu et al. 2005), and prostate cancer cells (Li et al. 2005). This has encouraged new research to target overexpressed γ-tubulin (Ebisu et al. 2021) for the development of new drugs to control proliferation of cancer cells. The cytotoxicity of gatastatin (O7-benzyl glaziovianin A derivative) was evaluated which showed some but weak cytotoxicity, while co-treatment with Plk1 inhibitor BI 2536 exhibited strong cytotoxicity and is being further evaluated for its therapeutic potential. The mechanism underlying the cytotoxicity was determined as spindle assembly checkpoint-dependent mitotic cell death by impairing centrosome functions (Ebisu et al. 2021). Further optimization of gatastatin revealed that the highly specific γ-tubulin inhibitor gatastatin 2 (O6-propargyl-gatastain) showed increased cytotoxicity, induced abnormal spindle formation with misaligned chromosomes, and inhibited microtubule nucleation (Shintani et al. 2020).

*Pericentrin* is an important centrosome protein that plays a critical role in recruiting γ-tubulin to centrosomes (Dictenberg et al. 1998), and it is part of the pericentrin/AKAP450 centrosomal targeting (PACT) domain (Gillingham and Munro 2000). It is critically involved in centrosome and spindle organization (Dictenberg et al. 1998; Doxsey et al. 1994; Young et al. 2000). Pericentrin forms ca. 3-MDa complex with γ-tubulin and depends on dynein for assembly onto centrosomes (Young et al. 2000). Mutation of the pericentrin gene results in loss of recruitment of several other centrosomal proteins that play a role in various diseases including ovarian and other cancers (reviewed in Badano et al. 2005; Schatten and Sun 2018; Morretton et al. 2022).

*Centrins* are members of a highly conserved subgroup of the EF-hand superfamily of $Ca^{2+}$-binding proteins, being primarily associated not only with centrioles but also with centrosomes to play an essential role in the duplication of centrosomes (Salisbury et al. 2002; Levy et al. 1996; Salisbury 1995; Lutz et al. 2001) (reviewed in Manandhar et al. 2005). It has been shown that pathogens like *Chlamydia* and certain bacteria can release certain components (type III-secreted effector protein CteG by *Chlamydia trachomatis*) that bind to centrin-2 (CTN2), interfering with accurate centrosome duplication, resulting in centrosome amplification as is characteristic for cancer (Steiert et al. 2023; Sauer et al. 2022); the effect persists even when the pathogen is removed. These findings explain why chlamydial infections are associated with an increased risk of cervical and ovarian cancer (Steiert et al. 2023) and delineates one of the increasing findings showing that centrosome abnormalities can be triggered by pathogens. Centrin-2 in this case is primarily affected leading to a cascade of subsequent centrosomal abnormalities.

*NuMA* (nuclear mitotic apparatus protein) is an intriguing multifunctional protein (reviewed in Sun and Schatten (2006, 2007)) that is essential for the precise organization of the mitotic apparatus during mitosis. In interphase, NuMA serves as nuclear matrix protein; it is not associated with the interphase centrosome. At the onset of mitosis, during nuclear envelope breakdown, NuMA becomes dispersed into the cytoplasm (Saredi et al. 1997) and becomes phosphorylated in a Cdk1-/cyclin-B-dependent process (Saredi et al. 1997), which allows its association with microtubules using a dynein–dynactin-mediated mechanism for its translocation to the centrosome area where it forms an insoluble crescent around centrosomes that tethers microtubules precisely into the bipolar mitotic apparatus (Merdes and Cleveland 1998); it cross-links spindle microtubules to organize and stabilize spindle poles from early mitosis to anaphase. Its dissociation from the centrosome during the exit from mitosis requires cdc1/cyclin B (Gehmlich et al. 2004). Failure of NuMA to relocate to the nucleus will result in cytoplasmic NuMA spots that organize abnormal microtubule asters (Gehmlich et al. 2004) and may contribute to mitotic abnormalities.

NuMA abnormalities have been shown for breast cancer (Kammerer et al. 2005; Moreno et al. 2019), which includes signaling abnormalities for NuMA's association with centrosomes and specifically includes a signaling function of cyclin B (reviewed in Sun and Schatten 2006), but other signaling molecules may play a role in different cancers. NuMA has been shown to respond to hormonal signaling in cervical cancer cells (Yam et al. 1998). Overexpression of NuMA has been shown in epithelial ovarian cancer (Brüning-Richardson et al. 2012).

Various structural centrosome abnormalities are observed in cancer tissue as presented in Figs. 3.3 and 3.4. Morphological differences in cancer cell centrosomes compared to centrosomes in noncancerous healthy cells include increased centrosome number, increased centrosome volume, excess of centrosomal material, supernumerary centrioles, abnormally oriented centrioles, and a host of imbalanced centrosomal proteins surrounding centrioles.

The causes for such centrosomal abnormalities can be complex and can have different origins for different tissues. It includes imbalances or dysfunctions of structural proteins, regulatory molecules, motor-related molecules, and others perhaps not yet known. In addition, basal bodies derived from dislocated primary cilia (Lingle and Salisbury 1999; Schatten et al. 2000) (discussed below) can form additional nucleation sites for microtubule organization, resulting in mitotic and division abnormalities; dysfunction or loss of primary cilia structure can result in signaling abnormalities with subsequent cascades of centrosomal disorganization (discussed below) (Lingle and Salisbury 1999, 2000; Schatten et al. 2000; Lingle et al. 2002).

## 3.3 Centrosome Regulation in Normal Cell Cycles

To understand centrosome dysregulation in ovarian cancer, it is important to understand centrosome regulation in normal cell cycles and determine the specific origins of centrosome dysfunctions in ovarian cancer. This knowledge will allow us to determine new targets for the development of ovarian cancer therapies to reverse and correct the dysregulated steps in the cascade leading to centrosome dysfunctions.

In a regular cell cycle, the centrosome cycle is well synchronized with the DNA/chromosome cycle to allow accurate separation of chromosomes during mitosis and cell division. Like DNA, it is important that centrosomes are precisely duplicated only once during the cell cycle to avoid abnormalities in mitosis and cell division. Several excellent studies have been conducted on the duplication of centrosomes and a block to reduplication has been established that assures accurate centrosome duplication only once within a normal cell cycle (reviewed in Schatten 2008; Schatten and Sun 2018; Fisk 2012; Prosser and Fry 2012; Fukasawa 2012; Boutros 2012).

The *centrosome duplication cycle* starts shortly before the G2 cell cycle stage in a precisely orchestrated duplication program. The duplication cycle begins with the disorientation of the pair of centrioles, centriole duplication, and centriole disjunction resulting in sister centriole separation (reviewed in Mack et al. 2000; Ou and Rattner 2004).

While we now understand the structural and molecular dynamics of centriole duplication, we are still in the process of completely understanding the duplication of the highly complex centrosomal matrix that has been correlated to duplication and separation of the better understood centriole cycle. While the mechanisms underlying centrosome matrix duplication are not yet fully understood, we do know that initiation of centrosome duplication is under cytoplasmic control and requires cyclin-dependent kinase 2 (Cdk2) complexed with cyclin E or cyclin A that rises during the late G1 stage (reviewed in Sluder 2004). The initiation of centrosome duplication further requires calcium/calmodulin-dependent kinase II (CaMKII) (Matsumoto and Maller 2002); CaMKII phosphorylates centrosome proteins in vitro (Pietromonaco et al. 1995) and it is localized to spindle poles (Ohta et al. 1990). Ubiquitin-mediated proteolysis of centrosomal proteins has been implicated in the block to reduplication, as a variety of components of the SCF (Skp1/cullin/F-box; ubiquitin ligase complex) proteolysis pathway as well as the 26S proteasome are localized to centrosomes in human cells (Fisk 2012; Prosser and Fry 2012; Fukasawa 2012; Kais and Parvin 2012; Tugendreich et al. 1995; Freed et al. 1999; Gstaiger et al. 1999).

*Centrosome separation* begins with the spatial separation of centrosome material around the nucleus and continues during mitosis which in part is driven by plus- and minus-end-directed microtubule motor proteins. Phosphorylation of the centrosomal protein centrin plays a role in centrosome disjunction at the G2/prophase transition, and it has been shown that the Nek2 kinase is involved in centrosome disjunction (Meraldi and Nigg 2001) (reviewed in Fry 2002).

For the *synchronization of centrosome duplication and DNA replication*, several molecular processes are common, which includes hyperphosphorylation of the retinoblastoma (RB) protein and activation of Cdk2, as well as cell cycle-dependent checkpoints. When the synchronization of the centrosome cycle and the DNA cycle becomes uncoupled, misregulation abnormalities become apparent which can lead to disease states, and it is especially apparent in cells during carcinogenesis (reviewed in Schatten 2008; Schatten and Sun 2018).

*Centrosome phosphorylation* is an important aspect for a regular centrosome cycle (reviewed in Fisk 2012; Prosser and Fry 2012; Fukasawa 2012; Boutros 2012) allowing specific centrosome functions such as increased microtubule nucleation and organization to enable the transition from G2 to mitosis which includes specific centrosome proteins that depend on multiple signaling. Accurate phosphorylation of centrosome proteins is important for the transition from G2 to mitosis (G2/M) in a process termed centrosome maturation which allows centrosomes to become division-competent and allow cell proliferation (reviewed in Schatten 2008). Cancer cell centrosomes become abnormally phosphorylated which may be related to misguided signal transduction resulting in mistargeting of key cell cycle regulators. As mentioned above, cyclin-dependent kinases are critically important for centrosome cycle progression. The G2/M transition requires Cdk1/cyclin B as well as Cdk1/cyclin A (reviewed in Fry and Hames 2004). Cdk1 is localized to centrosomes at the onset of mitosis (Bailly et al. 1989; Pockwinse et al. 1997), and Cdk1/cyclin B activation is detected in centrosomes during prophase (Jackman et al. 2003). Centrosome remodeling during G2/M includes the acquisition of polo-like kinase 1 (Plk1) (Golsteyn et al. 1995), NuMA (Merdes et al. 1996), and others, while interphase centrosome proteins such as C-Nap1 (Fry et al. 1998) or Nlp (Casenghi et al. 2003) are removed. γ-TuRC recruitment to the centrosome increases, assuring increased nucleation of microtubules for spindle formation.

Several critical mitotic cell cycle regulators are concentrated at the centrosome and include polo and Aurora kinase A (Barr and Gergely 2007) as well as cdc2/cyclin B kinase (Jackman et al. 2003). Several proteins that are concentrated at the centrosome bind to the anaphase promotion complex/cyclosome (APC/C). The activated APC/CCdc20 degrades cyclin B and securin to allow cell cycle exit from mitosis (Kramer et al. 2000; Huang and Raff 1999; Wakefield et al. 2000; Wei et al. 2011). Polo-like kinases are important for multiple stages of mitotic progression and include centrosome separation. Specific kinases including Plk1 are involved in centrosome and microtubule organization (Sun et al. 2001a, b, c, d, 2002; Tong et al. 2002, 2003; Fan et al. 2003); Plk1 and Plk3 both have been implicated in microtubule and centrosome functions in interphase and in mitosis (Fenton and Glover 1993; Donaldson et al. 2001; Wang et al. 2002). Loss of Plk3 function has also been associated with loss of cell shape (Wang et al. 2002), affecting microtubule functions beneath the plasma membrane, a factor that may play a role in loss of cellular polarity in cancer cells and tissue.

The composition of a functional mitotic centrosome requires microtubule motor proteins for translocation of several centrosomal proteins along microtubules to the centrosome core structure including translocation of pericentrin, centrin, ninein, and NuMA. Imbalanced or disrupted transport of these centrosome proteins has been implicated in pathologies related to centrosome and microtubule functions as well as to failures in organelle and vesicle distribution. Disruption of transport may either be the cause or effect in a chain of signal transduction events which can result in secondary pathologies.

## 3.4 Centrosome Dysregulation, Abnormalities in Ovarian Cancer, and New Research to Correct Centrosome Abnormalities

In recent years, several labs have focused on analyzing the cause and effect of centrosome dysfunctions in cancer cells with the goal to uncover and perhaps repair centrosomal pathologies on molecular levels. As centrosomal abnormalities in cancer cells can vary in different tissues, the origins of such abnormalities also can vary and require detailed tissue-specific analysis. For many cancers, a high number of mutation rates in cells pre-dispositioned to tumor may be the result of oncogenic insults, and the mutation rate increases in cells that have reached replicative senescence.

So far, excellent studies have been performed on cervical carcinogenesis that has strongly been associated with infections by high-risk human papillomavirus (HPVs). It has been proposed that the HPV E7 oncoprotein may induce primary centrosome duplication errors and act as mitotic mutator (reviewed in Münger and Duensing 2004). Abnormal multipolar mitoses resulting from supernumerary centrosomes have clearly been shown in HPV-associated lesions, and centrosome abnormalities are already detected in early stages of tumor development. It has been shown that uncoupling of the cell division cycle from the centrosome cycle subverts centrosome homeostasis (Duensing et al. 2000). Dissociation of centrosome cycles from DNA cycles may be among the causes for centrosome aberrations and multipolar spindle formations after irradiation. Aberrant hypermethylation has recently been implicated in inactivation of checkpoint genes that may influence cell cycle-dependent centrosome abnormalities as reported for pancreatic cancer (Ohki et al. 2000; Sato et al. 2003).

Environmental factors can cause significant damage to centrosomes, either directly or indirectly. It has been shown that environmental stress can result in the formation of aggresomes, which are oftentimes localized close to centrosomes and are thought to be the result of misfolded proteins (Ellgaard et al. 1999; Johnston et al. 1998; Wojcik and DeMartino 2003; Kopito 2000; Roth et al. 2008). Some of the aggresomes contain γ-tubulin and are associated with disease or disorders including noncancer diseases such as Parkinson's and dementia (McNaught et al. 2002).

In cancer cells, overexpression of specific centrosome proteins results in abnormal centrosome configurations and aneuploidy (Lingle et al. 2002; Katayama et al. 2003). Increased centrosome number and volume, supernumerary centrioles, accumulation of increased PCM, and abnormal phosphorylation of centrosomes have all been associated with cancer cell centrosomes, and in most cases followed by loss of cell polarity (Badano et al. 2005; Lingle et al. 1998). Several factors play a role in centrosome dysregulation, subsequent abnormal microtubule nucleation, abnormal spindle formation, and chromosomal missegregation, which includes the loss of tumor suppressor genes that affect centrosome functions.

Aurora kinases are among the critical kinases that play central roles in the mitotic process and cell division. Aurora kinase A (AURKA) is an important centrosome-associated serine/threonine kinase that has been shown to play a role in tumor development when overexpressed in animal models (Goepfert et al. 2002; Wang et al. 2006). Overexpression of the centrosomal protein Aurora-A results in multipolar mitotic spindles that play a role in early development of mammary tumors. It is associated with poor prognosis in epithelial ovarian cancer patients (Landen et al. 2007). Furthermore, Aurora kinase A expression predicts platinum resistance and adverse outcome in high-grade serous ovarian carcinoma patients (Mignogna et al. 2016). Therefore, targeting Aurora A is of high clinical significance.

Misregulation of centrosome duplication and genes implicated in centrosome amplification are among the factors involved in cell cycle-related abnormalities. The loss of the *tumor suppressor p53* is associated with multiple cycles of centrosome duplication in only one S phase resulting in multiple centrosome numbers (Pihan and Doxsey 1999), suggesting that p53 might play a role in the block to reduplication resulting in asynchrony with the DNA cycle. Viral oncoproteins that inactivate p53 also result in cells with supernumerary centrosomes as has been shown for the papillomavirus (reviewed in Münger and Duensing 2004). Different investigators, however, attribute centrosome abnormalities to missing checkpoint functions after loss of p53 rather than the loss of p53 directly affecting centrosome duplication (Meraldi et al. 2002). Other studies have shown that loss of p53 following genotoxic stress or mitogenic stimulation plays a role in the generation of supernumerary centrosomes in breast cancer cells in which the CDK2-/cyclin-dependent pathway is implicated (D'Assoro et al. 2004, 2008).

The breast- and ovary-specific *tumor suppressor gene BRCA1* has been implicated in the deregulation of centrosome duplication. During a regular cell cycle, BRCA1 is involved in G2/M checkpoint functions and it plays a role in preventing centrosome overduplication. Loss of BRCA1 may result in loss of the block to centrosome reduplication. Targeted deletion of BRCA1 exon11 leads to centrosome amplification (Xu et al. 1999); it causes centrosome overduplication and perhaps centrosome fragmentation in human breast cancer cell lines (Starita et al. 2004; Schlegel et al. 2003; Ko et al. 2006; reviewed in Chan 2011). While the exact mechanism by which BRCA1 affects centrosome duplication remains to be fully clarified, a model for the regulation of the centrosome by BRCA1 has been presented by Kais and Parvin (Kais and Parvin 2012).

In normal cell cycles, BRCA1 forms a complex with the BRCA1-associated RING domain 1 (BARD1) acting as E3 ubiquitin ligase. The BRCA1–BARD1 complex ubiquitinates γ-tubulin and maintains centrosome homeostasis, preventing abnormal duplication in normal cell cycles and abnormal microtubule nucleation by γ-tubulin. The BRCA1-associated centrosomal ninein-like protein (Nlp) is also involved in centrosome abnormalities. In transgenic mice, spontaneous breast tumorigenesis occurs when Nlp is overexpressed, perhaps mimicking BRCA1 loss (Shao et al. 2010).

The role of *BRCA2* in centrosome functions is less explored than BRCA1, but it is known that the BRCA2-associated protein NPM forms a complex with ROCK2 to maintain numerical centrosome integrity. Aberrant regulation of this protein may result in centrosome overduplication and fragmentation (Wang et al. 2011).

The *centrosomal kinase NEK2* (Never in mitosis gene A-related kinase 2) further plays a role in promoting centrosome accumulation, which has been shown for several cancers including breast (Hayward et al. 2004) and ovarian cancers (Liu et al. 2014; Das et al. 2013; Kokuryo et al. 2019; Khedkar et al. 2021). Along with other kinases like aurora kinases and polo-like kinases (Lee et al. 2014; Fry et al. 2012), NEK2 is overexpressed in ovarian cancer; it cooperates with oncogenic pathways to promote metastasis (Das et al. 2013), and it has further been shown that NEK2 is associated with drug resistance in ovarian cancer (Liu et al. 2014). Therefore, NEK2 has been proposed as a critical target with therapeutic potential to combat ovarian cancer cell proliferation, metastasis, and drug resistance (Hayward and Fry 2006; Marina and Saavedra 2014; Lin et al. 2016; Xia et al. 2015; Henise and Taunton 2011).

In a regular cell cycle, NEK2 is responsible for starting centrosome separation at the G2/M phase (Fry et al. 1998), while overexpression of NEK2 results in chromosome instability and aneuploidy in cancer cells (Hayward and Fry 2006; Zhou et al. 2013; Xia et al. 2015) in addition to activating several oncogenic pathways. Clinically, overexpression of NEK2 is correlated with rapid relapse and poor outcomes despite treatment strategies which may be related to its potent cell cycle functions and drug resistance. The chemotherapeutic resistance by NEK2 is related to its regulation of several genes including the ATP-binding cassette subfamily G (ABCG), aldehyde dehydrogenase 1 family, member A1 (ALDH1A1), and retinoid X receptor alpha (RXRA) (Lin et al. 2016; Yang et al. 2014). The resistance includes cancer drugs such as 5-fluorouracil, tamoxifen, trastuzumab, paclitaxel, and doxorubicin (Marina and Saavedra 2014; Lee and Gollahon 2013; Wen et al. 2016), in which drug efflux plays an essential role in increasing drug resistance which has been shown by NEK2 depletion that reduced drug efflux pump activity and inhibited drug resistance (Yang et al. 2014). Therefore, it has been proposed to use NEK2 inhibition as therapeutic option, although site effects would need to be evaluated, as NEK2 plays a critical role in normal cell cycles, being low in G1 phase cells but increasing steadily through S and G2 and peaking in late G2/M.

Progress has been made in recent years to find or design potent NEK2 inhibitors which includes propynamide 16 as an irreversible, cysteine-targeted inhibitor of NEK2 that inhibited cellular NEK2 without affecting the mitotic kinases, cyclin-dependent kinase 1 (CDK1), aurora B, or polo-like kinase 1 (PLK1) (Henise and Taunton 2011). As ovarian cancer is still one of the hardest to treat, the approach to design small molecule inhibitors to inactivate NEK2 appears highly promising.

### 3.4.1 Centrosome Clustering

Cancer cells have developed survival mechanisms that are not required for normal cells; it includes clustering of amplified centrosomal material during first mitosis to achieve bipolarity followed by division into two cells. As mentioned above, amplified centrosomes in cancer cells can either form tripolar or multipolar mitosis which results in unequal cell divisions in which some cells receive an abnormal number of oncogenes providing growth advantage while others undergo conditions having a different set of genes that will likely result in apoptosis or pyroptosis (reviewed in Rao et al. 2022). To achieve bipolarity in cells with supernumerary centrosomes, cancer cells have developed a mechanism by which these supernumerary centrosomes are clustered at the two poles, thereby mimicking bipolarity of normal cells although with amplified centrosomal material (Kwon et al. 2008; Leber et al. 2010; Krämer et al. 2011; Marthiens et al. 2012; Patel et al. 2018). Amplified clustered cancer cell centrosomes that form abnormal bipolar spindles are not easily discernible from regular bipolar spindles with non-amplified centrosomes compared to the obvious tripolar or multipolar mitotic spindles that are most commonly seen in cancer cells.

Although the detailed mechanisms underlying centrosome clustering into bipolar mitotic centrosomes are not yet fully understood, several components of the cytoskeleton may be involved. Experiments by Kwon et al. (2008) showed that the actin cytoskeleton plays a role in centrosome clustering. The role of microtubule motor proteins in centrosome clustering in somatic cancer and noncancer cells has been discussed by Krämer et al. (2012). Other studies performed in *Drosophila* cells have shown that a large number of proteins associated with a variety of cellular functions are required for efficient centrosome clustering (Gergely and Basto 2008; Kwon et al. 2008), which includes genes that play a role in spindle assembly checkpoint (SAC) functions. It also requires spindle tension to cluster supernumerary centrosomes into a bipolar mitotic apparatus. The studies by Kwon et al. (2008) have shown that SAC does not monitor extra centrosomes but that the time during which SAC monitors proper kinetochore–microtubule attachments gives supernumerary centrosomes time for clustering into a bipolar mitotic apparatus. Inhibition of SAC prevents clustering of supernumerary centrosomes. It was also suggested that microtubule-associated proteins (MAPs) are involved in centrosome clustering.

Several studies have shown that efficient centrosome clustering critically depends on HSET (kinesin-related protein, also now better known as KIFC1, a kinesin-like protein (kinesin motor)). KIFC1 is non-essential in normal somatic cells, therefore only affecting clustering into abnormal bipolar spindles without causing damaging site effects in noncancerous cells. KIFC1 depletion blocks centrosome clustering and promotes multipolar divisions (Kwon et al. 2008; Krämer et al. 2011; Marthiens et al. 2012). These studies showed that HSET (KIFC1) selectively eliminates cells with supernumerary centrosomes without affecting viability of cells with normal centrosome numbers. KIFC1 overexpression has been reported for several cancers including ovarian cancer, being associated with increased aggressiveness of the disease, poor prognosis, and drug resistance (De et al. 2009; Mittal et al. 2016, Pawar et al. 2014; Li et al. 2020; Hu et al. 2020). Preventing centrosome clustering into an abnormal bipolar mitotic apparatus may provide new targets for ovarian cancer therapy, as multipolar mitoses undergo fragmented cell divisions with fewer chances for cancer cell viability (Ganem et al. 2009) compared to abnormal bipolar mitotic spindles that can undergo cell division.

More research works are needed to possibly define optimal treatment concentrations and applications. Targeting KIFC1 has been a promising approach for breast cancer therapy, as KIFC1 is upregulated in breast cancer (Li et al. 2015), while it is absent in normal human mammary epithelial cells. Furthermore, KIFC1 has served as a biomarker in serous ovarian adenocarcinomas, indicating and predicting the aggressiveness of the disease (Mittal et al. 2016).

## 3.5 The Role of Primary Cilia in Ovarian Cancer

The primary cilium has gained increased attention especially during the past few decades when it was shown that it plays a critical role in essential cellular functions which includes an important role in signal transduction (reviewed in

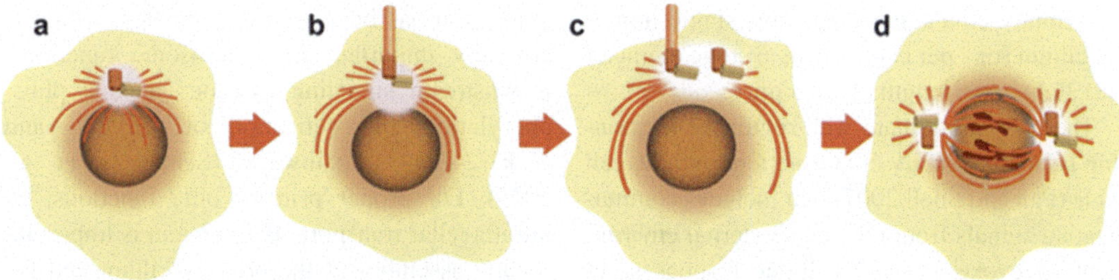


**Fig. 3.5** Schematic representation of primary cilium showing the close cell cycle relationships of the primary cilium with the cell's centrosome complex. In G1 (**a**) the mother centriole associates with membrane components to build the axoneme of the primary cilium. The primary cilium grows during G2 (**b**), duplicates during the S phase (**c**), and becomes disassembled during mitosis (**d**), when centrioles become associated again with the mitotic spindle poles. (Modified from Schatten (2014); reprinted from Schatten and Sun (2018))

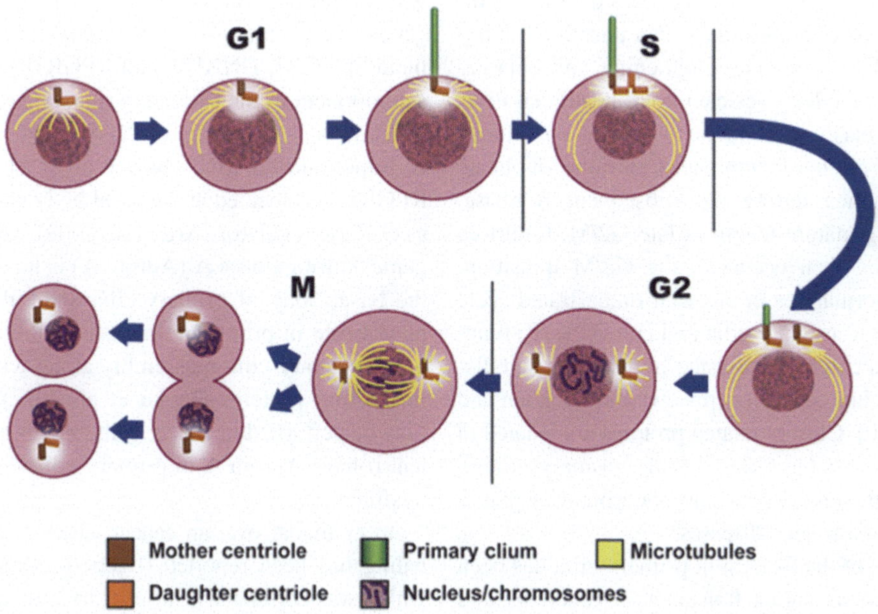


**Fig. 3.6** Schematic representation of primary cilium—cell cycle relationships and interactions with mitotic poles. The mother centriole undergoes docking to membrane in G1, while accessory proteins build the ciliary axoneme from the mother centriole's triplet microtubules. Cell centriole and DNA replication start during S, and centriole maturation occurs in G2. During mitosis centrioles participate in mitotic spindle formation. The primary cilium reassembles in G1. (Adapted from Schatten and Sun (2010); reprinted from Schatten and Sun (2011))

Badano et al. 2005; Schatten and Sun 2010) (Figs. 3.5 and 3.6).

The primary cilium is a specialized nonmotile single cilium protruding from almost all cells in our body (Wheatley et al. 1996; D'Angelo and Franco 2009; Veland et al. 2009). It is formed when dividing cells exit mitosis and the mother (elder) cell centriole becomes located at the cellular surface (Sorokin 1962), forming a basal body that nucleates microtubules that grow and build the interior skeleton of the primary cilium (reviewed in Schatten 2008; Schatten and Sun 2010, 2018).

The primary cilium is composed of a highly specialized receptor-rich plasma membrane with tissue-specific functions and an internal core consisting of nine outer microtubule doublets with no central microtubule pair ("9 + 0"), comprising the primary cilia's axoneme. This single immotile cilium (one per cell) has an average diameter of 0.2 μm and a length of ca. 8 μm.

There is an intimate well-regulated relationship between primary cilia functions and the cell cycle (Pan and Snell 2007) that includes communicating signals from the cell's external environment to evoke internal cellular responses. In noncancerous cells, the primary cilia cycle is precisely synchronized with the cell cycle and includes the centriole-centrosome cycle of its associated cell.

The primary cilium–centriole–centrosome cycle begins during G1 when the distal end of the mother centriole becomes associated with a membrane vesicle (reviewed by Pan and Snell 2007) followed by axoneme lengthening and enlargement of the ciliary vesicle into a sheath that fuses with the plasma membrane to become continuous with the plasma membrane. Centrioles duplicate and lengthen during the subsequent S phase, achieving mature length in late G2/M. Centriole shortening then occurs at the G2/M transition. Any abnormalities in this well-orchestrated cycle will lead to primary cilia and cell cycle dysfunctions that can be manifested in disorders and diseases including cancer (reviewed in Schatten and Sun 2010). Closely related proteins are located in all three structures (basal body, primary cilium, and centrosomes), and any abnormalities play a role in cilia-related diseases.

Much of the interest in primary cilia has been related to its signal transduction functions and the coordination with cell cycle regulation. Detailed signal transduction studies have revealed that the coordination of signal transduction cascades between primary cilia and the centrosome is essential for accurate cell cycle progression (reviewed in Schatten 2008; D'Angelo and Franco 2009; Veland et al. 2009; Quarmby and Parker 2005; Hildebrandt and Otto 2005; Davenport and Yoder 2005; Michaud and Yoder 2006; Satir and Christensen 2008). At least three important pathways require signaling through primary cilia and include the Wnt, hedgehog, and platelet-derived growth factor (PDGF) pathways (Sharma et al. 2008; Berbari et al. 2009). Furthermore, MAP kinase signaling through primary cilia has been well documented, which is critically important for centrosome functions. Downstream signaling cascades include phosphorylation and activation of the Akt and Mek1/2–Erk1/2 pathways (Schneider et al. 2005). For proper primary cilia functions, an intraflagellar transport (IFT) system is important for the assembly of the primary cilium and for transport of molecules to the cell body.

In cancer cells, the primary cilium becomes dislodged during progressive stages of cancer development (Lingle et al. 1998; Schatten et al. 2000; Lingle and Salisbury 2000), which coincides with the loss of cellular polarity. On genetic and molecular levels, several cilia-associated genes are commonly mutated in cancer cells including Gli1, DNAH9, and RPGR1P1. Gli1 is a component of the hedgehog signaling pathway; DNAH9 is a component of the dynein motor that is important for IFT (Bartoloni et al. 2001). RPGR1P1 is located at the basal body and serves as GTPase regulator (Arts et al. 2009). The oncogenic Aurora kinase A (Aurora A) is localized to the basal body of primary cilia and either may play a role in primary cilia disassembly or may block primary cilia reassembly along with other interacting proteins (Inoko et al. 2012), which may be perturbed in cancer cells, ascribing a critical role of Aurora A in primary cilia–cell cycle dysfunctions.

In epithelial ovarian cancer, aberrant cell signaling has been reported (Egeberg et al. 2012), which was correlated with primary cilia dysfunctions. In this study, using primary cultures of human OSE (ovarian surface epithelial) cells as well as cultures of human ovarian cancer cell lines, defects in the primary cilium were implicated in tumorigenesis, mainly due to deregulation of ciliary signaling pathways such as Hedgehog (Hh) signaling. The study showed that ovarian cancer cells display significantly reduced numbers of primary cilia, thereby affecting Hh signaling.

The role of primary cilia in hedgehog signaling is one of the best studied signaling pathways in noncancerous cells in which localization of the transmembrane protein, Smoothened (Smo), in the primary cilium plays a role in the activation of the hedgehog pathway; hedgehog-dependent transcription is mediated by the three transcription factors, Gli1, Gli2, and Gli3 (Kasper et al. 2006; Haycraft et al. 2005).

The study by Egeberg et al. (Egeberg et al. 2012) further shows that in addition to deregulated Hh signaling, ovarian cancer cells have deregulated platelet-derived growth factor receptor alpha (PDGFRα) expression.

The regulation of the PDGF signaling pathway by primary cilia has been well explored by Schneider et al. (Schneider et al. 2005) who showed that the platelet-derived growth factor receptor-α (PDGFRα) localizes to primary cilia in murine embryonic fibroblast cells; it was further shown that activation of Akt and ERK1/2 pathways by PDGF ligand requires primary cilia. Expression of PDGFRα is a poor prognostic indicator of breast cancer (Jechlinger et al. 2006; Carvalho et al. 2005), clearly linking PDGF signaling through primary cilia to breast cancer, which may also be the case for other cancers including ovarian cancer.

In ovarian cancer, it has been shown that primary cilia play a role in maintaining ovarian surface epithelium (OSE) homeostasis, while the low number of primary cilia in cancer OSE cells may result in part to be the result of overexpression of AURA, then resulting in aberrant Hh signaling leading to ovarian tumorigenesis (Egeberg et al. 2012).

As Aurora A is highly overexpressed in ovarian cancer as well as in other cancers (Tanner et al. 2000; Reiter et al. 2006; Shang et al. 2009; Jeng et al. 2004) and associates with centrosomes and primary cilia, it has been proposed that Aurora A plays a critical role in the disassembly of primary cilia upon mitogenic stimulation (Pugacheva et al. 2007). This area of research is very active and new data are being generated to determine the relationship between loss of primary cilia and their causative or consequential relationship during ovarian tumorigenesis. The development of agents targeting Hh pathway components, PDGFR, and AURA are actively being explored to better manage ovarian cancer (Campos and Ghosh 2010).

A higher expression level of the alpha form of PDGFR (PDGFRα) in ovarian cancer cells compared to normal ovarian surface epithelium (OSE) cells has been shown to be associated with high tumor grade, high proliferation index, and poor survival rates (Apte et al. 2004; Henriksen et al. 1993; Lassus et al. 2004; Matei et al. 2006).

AURA is a critical mitotic kinase involved in centrosome maturation, mitotic entry, and spindle assembly (Gillingham and Munro 2000). AURA is amplified in human ovarian cancer (Yan and Chng 2012; Olivero 2012; Salisbury et al. 2002), and elevated AURA kinase activity and/or increased protein levels are seen in ovarian cancer (Yan and Chng 2012; Olivero 2012; Salisbury 2004; Kawaguchi and Zheng 2004). More research works are still needed to better understand the multiple signaling pathways associated with ovarian tumorigenesis in order to identify new targets in the signaling pathways of EOC to be able to develop better treatment strategies.

## 3.6 Centrosomes as Target for Ovarian Cancer Therapy

Excellent general as well as detailed ovarian cancer treatment information for patients and health professionals is available from the National Institutes of Health and from the American Cancer Society, which includes new information on clinical trials that addresses many specific aspects that will not be addressed in this chapter.

Various treatment options are available that are tailored to individual tumors and include surgery, hormone therapy, targeted drug therapy, and immunotherapy (reviewed in Arora et al. 2023). Significant progress has been made in diagnosing ovarian cancer with various imaging technologies as well as in treatment strategies of ovarian cancer, which has led to personalized treatment procedures that target different abnormalities based on different subtypes of cancer and different dysfunctions of molecular pathways and molecular abnormalities.

Ovarian cancer treatment can be complex owing in part to the heterogeneity of the tissue and the fact that ovarian cancer is typically diagnosed at late stages of the disease. Tissue heterogeneity and genomic instability are among the complexities associated with ovarian cancer that require detailed histological analysis of the individual cancer in order to design effective treatment that may require multiple targeted treatment strategies. Centrosomes are increasingly being discussed as new targets for ovarian cancer treatment, as abnormal centrosomes are associated with abnormal cell divisions that are hallmark characteristics for cancer cells and tissue.

Targeting cancer cell centrosomes includes targeting abnormal centrosomes directly or targeting signal transduction molecules that play a role in abnormal centrosome formation. Such approaches may target overexpressed centrosome proteins, abnormal centrosome clustering, abnormal primary cilia dynamics, overexpressed phosphorylation such as Aurora A that is implicated in centrosome hyperphosphorylation or other components in the phosphorylation cascade, as well as different molecules that play a role in centrosome function such as the aryl hydrocarbon receptor (AhR) and cyclin E, as reported by Korzeniewski et al. (2010).

Pharmaceutical intervention known as chemotherapy is most commonly used after surgery and/or radiation, depending on the stage of the cancer and severity. Excellent information on treatment options is available from the websites of the National Institutes of Health (NIH), the American Cancer Society (ACS), the Center for Disease Control (CDC), and others that provide details on chemotherapy, hormone therapy, targeted drug therapy, immunotherapy, clinical trials, dietary considerations, and others. A recent review has been published with excellent details on ovarian cancer and current treatment strategies (Arora et al. 2023).

As mentioned above, like several other cancers, ovarian cancer displays significant heterogeneities, referring to cells that are different from each other which may require combination therapies to eradicate cells with different etiological abnormalities that may have been diagnosed for specific subpopulations within the ovarian cancer tissue. In the following, the focus will be on pharmaceutical intervention and future therapy potential as it concerns centrosomes and the microtubule cytoskeleton.

One of the best-known plant derivatives that have been developed into a cancer-targeting pharmaceutical is Taxol and its derivatives such as paclitaxel and dodetaxel. Paclitaxel (or Taxol) had originally been isolated from the gymnosperm *Taxus brevifolia*. It is best known for its targeting of microtubule dynamics by primarily inhibiting microtubule depolymerization, thereby preventing dynamic progression of cells to undergo mitosis and cell division (Schiff et al. 1979; Schatten et al. 1982a, b; others). Interactions of Taxol with microtubules at the centrosome–microtubule nucleation sites have been reported (De Brabander et al. 1981; Dimitriadis et al. 2001), and it had also been proposed that centrosomes in Taxol-treated cells may lose their capacity to nucleate microtubules (De Brabander et al. 1981); furthermore, Taxol may have other molecular target interactions that have not yet been explored fully. Taxanes and epithilones are widely used for epithelial ovarian cancer (EOC) (Tymon-Rosario et al. 2021) as well as breast cancer. However, its limited applications include drug resistance that typically develops within a certain time (reviewed in Reader et al. 2022; Barlow et al. 2002; Yin et al. 2010); active research is underway to overcome drug resistance and pursue new drug developments aimed at targeting the mitotic molecular machinery to prevent mitosis progression or to induce mitotic cell death to inhibit cancer cell division.

Other plant-derived drugs are widely used to control a variety of diseases and include the microtubule drug colcemid and various others. However, unlike colcemid or especially Taxol that has been thoroughly investigated and modified for the use as therapeutic anticancer drug, other plant-derived components have not yet undergone similar testing and are not yet available for efficient ovarian cancer treatment. This includes the potential use of curcumin, a natural polyphenol found in the rhizomes of *Curcuma longa* (turmeric) that has been shown to prevent or inhibit abnormal signaling cascades in cancer cells (Safavy et al. 2007; Yoon and Liu 2007; Yallapu et al. 2010a, b; Dev et al. 2012), which may include abnormal signaling to centrosomes.

Curcumin and curcumin derivatives show potent inhibitory effects on several signaling pathways including the NFκB pathway (Yamaguchi et al. 2012) that plays a role in inflammation related to cancer development and progression. The effects of curcumin on the NFκB pathway may affect centrosome dynamics. Several other pathways have been implicated (Wilken et al. 2011) and epigenetic effects have also been proposed (Reuter et al. 2011).

Curcumin has gained increased attention as potent anti-ovarian cancer drug, and several studies have reported that curcumin can delay ovarian cancer and increase its sensitivity to chemotherapy, as well as being able to reduce chemotherapy drugs' side effects (Mohamadian et al. 2022) and enhance the anticancer efficacy by paclitaxel in ovarian cancer (Liu et al. 2023). Because of its excellent potential for its use in ovarian cancer therapy, new more potent curcumin analogues are now being developed to increase its antitumor efficiency (Nagaraju et al. 2012; Yamaguchi et al. 2012), and further modifications are being pursued to increase its solubility and stability and other factors impacting its bioavailability for more potent metabolic effects. These include encapsulation of curcumin into liposomes, polymeric nanoparticles, biodegradable microspheres, curcumin–cyclodextrin complexes, and micellar formulations and hydrogels (Yallapu et al. 2010a, b; Anand et al. 2007; Tonnesen et al. 2002; Wang et al. 2008; Mukerjee and Viswanatha 2009; Cartiera et al. 2009). Other, still more promising strategies have also been proposed, including the use of theragnostic curcumin-encapsulated nanoparticles that will increase bioavailability and allow more potent clinical applications (Shin et al. 2012). Further specific improvements to increase bioavailability and stability include use of didodecyl dimethyl ammonium bromide (DDAB) and pluronic F127 polymer as a surfactant/stabilizer, in association with the use of a rotating tube microfluidic platform (Dev et al. 2012). Such improvements are important and highly promising to modify the dietary ingredient into a potent anti-ovarian cancer pharmaceutical.

The antimitotic drug *griseofulvin* has also shown great promise to be used as anticancer drug, as it affects several mechanisms related to centrosome dynamics. Griseofulvin arrests cells at the G2/M transition stage in a concentration-dependent manner and it may provide indications that the NFκB pathway and centrosome dynamics are connected, as griseofulvin has been shown to interfere with the NFκB pathway (Uen et al. 2007). Quite importantly, griseofulvin has clearly been shown to specifically inhibit supernumerary centrosome clustering in cancer cells which demonstrates its potent effects on centrosome dynamics (reviewed in Krämer et al. 2012) and inhibition of abnormal centrosome functions.

As mentioned above, prevention of amplified centrosome clustering into a bipolar mitotic apparatus is an important approach to prevent abnormal cancer cell mitosis. It can be used as therapeutic targeting to control ovarian cancer cell proliferation, as such centrosome clustering is not needed for normal cell mitosis. Clustering pathways are dispensable in cells with normal centrosome numbers, but centrosome clustering is required for supernumerary centrosomes to be distributed into the bipolar mitotic apparatus. In cancer cells, inhibiting centrosome clustering will prevent abnormal mitosis, as non-clustered centrosomes cannot form an abnormal mitotic apparatus but induce cell fragmentation and cell death rather than allowing formation of aneuploid cells with consequences for genomic instability.

Griseofulvin has been identified as potent inhibitor of centrosomal clustering (Rebacz et al. 2007) and it may further affect multiple sites in the mitotic process. Griseofulvin is already approved as an effective orally administered antifungal drug that affects microtubule formation and disrupts microtubule dynamics in vivo and in vitro (Marchetti et al. 1996; Schatten 1977; Wehland et al. 1977; Schatten et al. 1982a, b), which has been shown in several cell systems (Schatten 1977; Schatten et al. 1982a, b; Grisham et al. 1973; Miao et al. 2012; other references therein). Several investigators have shown that griseofulvin induces multipolar mitoses in tumor cells (Rebacz et al. 2007; Schatten 1977; Ho et al.

2001; Panda et al. 2005); detailed mechanisms underlying its inhibition of centrosome clustering are still being explored. Griseofulvin may interfere with microtubule minus ends rather than interacting with the centrosome structure directly. Its action as antitumor drug needs further investigation, as determining its molecular mechanisms would allow specific targeting of ovarian cancer subpopulations and further allow its use in combination with other drugs to allow for more effective actions in combination therapies.

## 3.7 Conclusion and Future Directions

Significant progress has been made in the overall diagnosis and treatment of ovarian cancer including personalized treatment procedures that target specific abnormalities in specific types of ovarian cancer. However, despite the progress of being able to administer effective drugs, drug resistance is still a major problem and new drug developments are needed to overcome the current shortfalls. During the past few years, renewed interest has focused on centrosomes as target for ovarian cancer therapy, as centrosomes play crucial roles in allowing abnormal cell divisions that are hallmarks for cancer cells and tissue.

Unlike centrosomes in noncancer cells, ovarian cancer cell centrosomes are dysregulated, owing to several consequential abnormalities including abnormal phosphorylation and supernumerary centrosome formations that can cluster to form abnormal mitoses and cause several other dysfunctions that are characteristic of cancer cell centrosomes (reviewed in Schatten and Sun 2018; Schatten 2022a, b). Dysregulation of centrosomes leads to centrosome amplification with consequences for genomic instability and loss of tumor suppressor genes. Several new studies have started to determine mechanisms to control centrosome amplification, which includes inhibiting signaling pathways that allow centrosome amplification, some of which are tissue specific. Targeted destruction of centrosome clustering has been another approach to produce fragmented cells that will lose viability and not participate in abnormal cell divisions.

Other approaches include the development of new drug delivery systems to achieve more efficient bioavailability for more potent metabolic impact to destroy cancer cells. Such new developments are highly promising avenues for targeting ovarian cancer cell centrosomes and prevent or inhibit ovarian cancer cell growth and development into metastasis.

## References

Anand P, Kunnumakkara AB, Newman RA, Aggarwal BB (2007). Bioavailability of curcumin: problems and promises. Mol Pharm 4:807–818.

Andersen JS, Wilkinson CJ, Mayor T, Mortensen P, Nigg EA, Mann M (2003). Proteomic characterization of the human centrosome by protein correlation profiling. Nature 426:570–574.

Apte SM, Bucana CD, Killion JJ, Gershenson DM, Fidler IJ (2004). Expression of platelet-derived growth factor and activated receptor in clinical specimens of epithelial ovarian cancer and ovarian carcinoma cell lines. Gynecol Oncol 93:78–86.

Arora T, Mullangi S, Lekkala MR (2023). Ovarian Cancer. In: StatPearls [Internet]. Treasure Island (FL): StatPearls Publishing; PMID: 33620837.

Arts HH, Cremers FP, Knoers NV, Roepman R (2009). Focus on molecules: RPGRIP1. Exp Eye Res 88:332–333.

Badano JL, Teslovich TM, Katsanis N (2005). The centrosome in human genetic disease. Nat Rev Genet 6:194–205.

Bailly E, Dorée M, Nurse P, Bornens M (1989). P34cdc2 is located in both nucleus and cytoplasm; part is centrosomally associated at G2/M and enters vesicles at anaphase. EMBO J 8:3985–3995.

Barlow JB, Gonzalez-Garay ML, Cabral F (2002). Paclitaxel-dependent mutants have severely reduced microtubule assembly and reduced tubulin synthesis. J Cell Sci 115:3469–3478.

Barr AR, Gergely F (2007). Aurora A: the maker and breaker of spindle poles. J Cell Sci 120:2987–2996.

Bartoloni L, Blouin JL, Maiti AK, et al (2001). Axonemal beta heavy chain dynein DNAH9: cDNA sequence, genomic structure, and investigation of its role in primary ciliary dyskinesia. Genomics 72:21–33.

Berbari NF, O'Connor AK, Haycraft CJ, Yoder BK (2009). The primary cilium as a complex signaling center. Curr Biol 19:R526–535.

Boutros R (2012). Regulation of centrosomes by cyclin-dependent kinases. In: Schatten H, editor. The centrosome, Chap 11. New York: Springer Science and Business Media.

Boveri T. (1914). Zur Frage der Entstehung maligner Tumoren. Jena: G. Fisher.

Boveri T (2008). Concerning the origin of malignant tumours by Theodor Boveri (Translated and annotated by Henry Harris). J Cell Sci. 2008;121 Suppl 1:1–84.

Brüning-Richardson A, Bond J, Alsiary R, Richardson J, Cairns DA, McCormac L, Hutson R, Burns PA, Wilkinson N, Hall GD, Morrison EE, Bell SM (2012). NuMA overexpression in epithelial ovarian cancer. PLoS One 7(6):e38945. https://doi.org/10.1371/journal.pone.0038945. PMID: 22719996; PMCID: PMC3375276.

Campos SM, Ghosh S (2010). A current review of targeted therapeutics for ovarian cancer. J Oncol 201:149362.

Cartiera MS, Ferreira EC, Caputo C, Egan ME, Caplan MJ, Saltzman WM (2009). Partial correction of cystic fibrosis defects with PLGA nanoparticles encapsulating curcumin. Mol Pharm 7(1):86–93.

Carvalho I, Milanezi F, Martins A, Reis RM, Schmitt F (2005). Overexpression of platelet-derived growth factor receptor alpha in breast cancer is associated with tumour progression. Breast Cancer Res 7:R788–795.

Casenghi M, Meraldi P, Weinhart U, Duncan PI, Korner R, Nigg EA (2003). Polo-like kinase 1 regulates Nlp, a centrosome protein involved in microtubule nucleation. Dev Cell 5:113–125

Chan JY (2011). A clinical overview of centrosome amplification in human cancers. Int J Biol Sci 7(8):1122–1144.

D'Angelo A, Franco B (2009). The dynamic cilium in human diseases. Patho Genetics 2(3):1–15.

Das TK, Dana D, Paroly SS, Perumal SK, Singh S, Jhun H, Pendse J, Cagan RL, Talele TT, Kumar S (2013). Centrosomal kinase Nek2 cooperates with oncogenic pathways to promote metastasis. Oncogenesis 2, e69; https://doi.org/10.1038/oncsis.2013.34

D'Assoro AB, Busby R, Suino K, et al (2004). Genotoxic stress leads to centrosome amplification in breast cancer cell lines that have an inactive G1/S cell cycle checkpoint. Oncogene 23:4068–4075.

D'Assoro AB, Busby R, Acu ID, et al (2008). Impaired p53 function leads to centrosome amplification, acquired ER alpha phenotypic heterogeneity and distant metastases in breast cancer MCF-7 xenografts. Oncogene 27:3901–3911.

Davenport JR, Yoder BK (2005). An incredible decade for the primary cilium: a look at a once forgotten organelle. Am J Physiol Renal Physiol 289:F1159–1169.

De Brabander M, Geuens G, Nuydens R, Willebrords R, De Mey J (1981). Taxol induces the assembly of free microtubules in living cells and blocks the organizing capacity of the centrosomes and kinetochores. Proc Natl Acad Sci USA 78:5608–12.

De S, Cipriano R, Jackson MW, Stark G R (2009). Overexpression of kinesins mediates docetaxel resistance in breast cancer cells. Cancer Res 69(20):8035–8042.

Dev S, Prabhakaran P, Filgueira L, Iyer KS, Raston CL (2012). Microfluidic fabrication of cationic curcumin nanoparticles as an anti-cancer agent. Nanoscale 4:2575–2579.

Dictenberg J, Zimmerman W, Sparks C, Young A, Vidair C, Zheng Y, Carrington W, Fay F, Doxsey SJ (1998). Pericentrin and gamma tubulin form a protein complex and are organized into a novel lattice at the centrosome. J Cell Biol 141:163–174.

Dimitriadis I, Katsaros C, Galatis B (2001). The effect of taxol on centrosome function and microtu bule organization in apical cells of Sphacelaria rigidula (Phaeophyceae). Phycol Res 49:23–34.

Donaldson MM, Tavares AAM, Hagan IM, Nigg EA, Glover DM (2001). The mitotic roles of polo like kinase. J Cell Sci 114:2357–2358.

Doxsey SJ, Stein P, Evans L, Calarco P, Kirschner M (1994). Pericentrin, a highly conserved protein of centrosomes involved in microtubule organization. Cell 76:639–650.

Dráber P, Dráberová E (2021). Dysregulation of Microtubule Nucleating Proteins in Cancer Cells. Cancers (Basel) 13(22):5638. https://doi.org/10.3390/cancers13225638. PMID: 34830792; PMCID: PMC8616210.

Duensing S, Lee LY, Duensing A, Basile J, Piboonniyom S, Gonzalez S, Crum CP, Munger K (2000). The human papillomavirus type 16 E6 and E7 oncoproteins cooperate to induce mitotic defects and genomic instability by uncoupling centrosome duplication from the cell division cycle. Proc Natl Acad Sci USA 97:10002–10007.

Ebisu H, Shintani K, Chinen T, Nagumo Y, Shioda S, Hatanaka T, Sakakura A, Hayakawa I, Kigoshi H, Usui T (2021) Dual Inhibition of γ-Tubulin and Plk1 Induces Mitotic Cell Death. Front Pharmacol 11:620185. https://doi.org/10.3389/fphar.2020.620185

Egeberg DL, Lethan M, Manguso R, Schneider L, Awan A, Jørgensen TS, Byskov AG, Pedersen LB, Christensen ST (2012). Primary cilia and aberrant cell signaling in epithelial ovarian cancer. Cilia 1:15. https://doi.org/10.1186/2046-2530-1-15

Ellgaard L, Molinari M, Helenius A (1999). Setting the standards: quality control in the secretory pathway. Science 286:1882–1888.

Fan H-Y, Tong C, Teng C-B, Lian L, Li S-W, Yang Z-M, Chen D-Y, Schatten H, Sun Q-Y (2003). Characterization of Polo-like kinase-1 in rat oocytes and early embryos implies its functional roles in the regulation of meiotic maturation, fertilization and cleavage. Mol Reprod Dev 65:318–329.

Fenton B, Glover DM (1993). A conserved mitotic kinase active at late anaphase-telophase in syncytial Drosophila embryos. Nature 363:637–640.

Fisk HA (2012). Many pathways to destruction: the centrosome and its control by and role in regulated proteolysis. In: Schatten H, editor. The centrosome, Chap 8. New York: Springer Science and Business Media.

Flory MR, Davis TN (2003). The centrosomal proteins pericentrin and kendrin are encoded by alter natively spliced products of one gene. Genomics 82:401–405.

Freed E, Lacey KR, Huie P, Lyapina SA, Deshaies RJ, Stearns T, Jackson PK (1999). Components of an SCF ubiquitin ligase localize to the centrosome and regulate the centrosome duplication cycle. Genes Dev 13:2242–2257.

Fry AM, Mayor T, Meraldi P, Stierhof YD, Tanaka K, Nigg EA (1998). C-NAP1, a novel centrosomal coiled-coil protein and candidate substrate of the cell cycle-regulated protein kinase Nek2. J Cell Biol 141:1563–1574. PMID: 2133000

Fry AM (2002). The Nek2 protein kinase: a novel regulator of centrosome structure. Oncogene 21:6184–6194.

Fry AM, Hames RS (2004). The role of the centrosome in cell cycle progression. In: Nigg E, editor. Centrosomes in development and disease. Weinheim: Wiley-VCA; p. 143–66.

Fry AM, O'Regan L, Sabir SR, Bayliss R (2012). Cell cycle regulation by the NEK family of protein kinases. J Cell Sci 125:4423–4433. PMID: 3500863. https://doi.org/10.1242/jcs.111195

Fry AM, Sampson J, Shak C, Shackleton S (2017). Recent advances in pericentriolar material organization: Ordered layers and scaffolding gels. F1000Research 6, 1622.

Fukasawa K (2012). Molecular links between centrosome duplication and other cell cycle associated events. In: Schatten H, editor. The centrosome, Chap 10. New York: Springer Science and Business Media.

Ganem NJ, Godinho SA, Pellman D (2009). A mechanism linking extra centrosomes to chromosomal instability. Nature 460(7252):278–82.

Gehmlich KL, Haren L, Merdes A (2004). Cyclin B degradation leads to NuMA release from dynein/dynactin and from spindle poles. EMBO Rep 5:97–103.

Gergely F, Basto R (2008). Multiple centrosomes: together they stand, divided they fall. Genes Dev 2008(22):2291–2296.

Gillingham AK, Munro S (2000). The PACT domain, a conserved centrosomal targeting motif in the coiled-coil proteins AKAP450 and pericentrin. EMBO Rep 1:524–529.

Goepfert TM, Adigun YE, Zhong L, et al (2002). Centrosome amplification and overexpression of aurora A are early events in rat mammary carcinogenesis. Cancer Res 62:4115–4122.

Golsteyn RM, Mundt KE, Fry AM, Nigg EA (1995). Cell cycle regulation of the activity and subcellular localization of Plk1, a human protein kinase implicated in mitotic spindle function. J Cell Biol 129:1617–1628.

Grisham LM, Wilson L, Bensch KG (1973). Antimitotic action of griseofulvin does not involve disruption of microtubules. Nature 244:294–296.

Gstaiger M, Marti A, Krek W (1999). Association of human SCF(SKP2) subunit p19(SKP1) with interphase centrosomes and mitotic spindle poles. Exp Cell Res 247:554–562.

Hannak E, Oegema K, Kirkham M, Gonczy P, Habermann B, Hyman AA (2002). The kinetically dominant assembly pathway for centrosomal asters in Caenorhabditis elegans is γ-tubulin dependent. J Cell Biol 157:591–602.

Haycraft CJ, Banizs B, Aydin-Son Y, et al (2005). Gli2 and gli3 localize to cilia and require the intraflagellar transport protein polaris for processing and function. PLoS Genet 1:e53.

Hayward DG, Clarke RB, Faragher AJ, et al (2004). The centrosomal kinase Nek2 displays elevated levels of protein expression in human breast cancer. Cancer Res 64:7370–7376.

Hayward DG, Fry AM (2006). NEK2 kinase in chromosome instability and cancer. Cancer Lett 237:155–166. PMID:16084011. https://doi.org/10.1016/j.canlet.2005.06.017.

Henise JC, Taunton J (2011). Irreversible NEK2 kinase inhibitors with cellular activity. J Med Chem 54:4133–4146. PMID:3663048. https://doi.org/10.1021/jm200222m

Henriksen R, Funa K, Wilander E, Backstrom T, Ridderheim M, Oberg K (1993). Expression and prognostic significance of platelet-derived growth factor and its receptors in epithelial ovarian neoplasms. Cancer Res 53:4550–4554.

Hildebrandt F, Otto E (2005). Cilia and centrosomes: a unifying pathogenic concept for cystic kidney disease? Nat Rev Genet 6:928–940.

Ho YS, Duh JS, Jeng JH, Wang YJ, Liang YC, Lin CH, Tseng CJ, Yu CF, Chen RJ, Lin JK (2001). Griseofulvin potentiates antitumorigenesis effects of nocodazole through induction of apoptosis and G2/M cell cycle arrest in human colorectal cancer cells. Int J Cancer 91:393–401.

Hsu LC, Kapali M, DeLoia JA, Gallion HH (2005). Centrosome abnormalities in ovarian cancer. Int J Cancer 113:746–751.

Hu Y, Zheng M, Wang C, Wang S, Gou R, Liu O, Li X, Liu J, Lin B (2020). Identification of KIF23 as a prognostic signature for ovarian cancer based on large-scale sampling and clinical validation. Am J Transl Res 12(9):4955–4976. PMID: 33042400; PMCID: PMC7540138.

Huang J, Raff JW (1999). The disappearance of cyclin B at the end of mitosis is regulated spatially in Drosophila cells. EMBO J 18:2184–2195.

Inoko A, Matsuyama M, Goto H, Ohmuro-Matsuyama Y, Hayashi Y, Enomoto M, Ibi M, Urano T, Yonemura S, Kiyono T, Izawa I, Inagaki M (2012). Trichoplein and Aurora A block aberrant primary cilia assembly in proliferating cells. J Cell Biol 197(3):391–405.

Jackman M, Lindon C, Nigg E, Pines J (2003). Active cyclin B1-Cdk1 first appears on centrosomes in prophase. Nat Cell Biol 5:143–148.

Jayson GC, Kohn EC, Kitchener HC, Ledermann JA (2014). Ovarian cancer. Lancet. 384(9951):1376–1388.

Jechlinger M, Sommer A, Moriggl R, et al (2006). Autocrine PDGFR signaling promotes mammary cancer metastasis. J Clin Invest 116:1561–1570.

Jeng YM, Peng SY, Lin CY, Hsu HC (2004). Overexpression and amplification of Aurora-A in hepatocellular carcinoma. Clin Cancer Res 10:2065–2071.

Johnston JA, Ward CL, Kopito RR (1998). Aggresomes: a cellular response to misfolded proteins. J Cell Biol 143:1883–1898.

Kais Z, Parvin JD (2012). Centrosome regulation and breast cancer. In: Schatten H, editor. The centrosome, Chap 14. New York: Springer Science and Business Media.

Kammerer S, Roth RB, Hoyal CR, Reneland R, Marnellos G, Kiechle M, Schwarz-Boeger U, Griffiths LR, Ebner F, Rehbock J, Cantor CR, Nelson MR, Brown A (2005). Association of the NuMA region on chromosome 11q13 with breast cancer susceptibility. Proc Natl Acad Sci USA 102(6):2004–2009.

Kasper M, Regl G, Frischauf AM, Aberger F (2006). GLI transcription factors: mediators of onco genic Hedgehog signalling. Eur J Cancer 42:437–445.

Katayama H, Brinkley WR, Sen S (2003). The Aurora kinases: role in cell transformation and tumorigenesis. Cancer Metastasis Rev 22:451–464.

Kawaguchi S, Zheng Y (2004). Characterization of a Drosophila centrosome protein CP309 that shares homology with Kendrin and CG-NAP. Mol Biol Cell 15:37–45.

Keryer G, Di Fiore B, Celati C, Lechtreck KF, Mogensen M, Delouvee A, Lavia P, Bornens M, Tassin AM (2003). Part of Ran is associated with AKAP450 at the centrosome: involvement in microtubule-organizing activity. Mol Biol Cell 14:4260–4271.

Khedkar HN, Wang YC, Yadav VK, Srivastava P, Lawal B, Mokgautsi N, Sumitra MR, Wu ATH, Huang HS (2021). In-Silico Evaluation of Genetic Alterations in Ovarian Carcinoma and Therapeutic Efficacy of NSC777201, as a Novel Multi-Target Agent for TTK, NEK2, and CDK1. Int J Mol Sci 22(11):5895. https://doi.org/10.3390/ijms22115895. PMID: 34072728; PMCID: PMC8198179.

Ko MJ, Murata K, Hwang DS, et al (2006). Inhibition of BRCA1 in breast cell lines causes the centrosome duplication cycle to be disconnected from the cell cycle. Oncogene 25:298–303.

Kokuryo T, Yokoyama Y, Yamaguchi J, Tsunoda N, Ebata T, Nagino M (2019). NEK2 Is an Effective Target for Cancer Therapy With Potential to Induce Regression of Multiple Human Malignancies. Anticancer Res 39(5):2251–2258. https://doi.org/10.21873/anticanres.13341. PMID: 31092416.

Kopito RR (2000). Aggresomes, inclusion bodies and protein aggregation. Trends Cell Biol 10:524–530.

Korzeniewski N, Wheeler S, Chatterjee P, et al (2010). A novel role of the aryl hydrocarbon receptor (AhR) in centrosome amplification—implications for chemoprevention. Mol Cancer 9:153.

Korzeniewski N, Duensing S (2012). Disruption of centrosome duplication control and induction of mitotic instability by the high-risk human papillomavirus oncoproteins E6 and E7. In: Schatten H, editor. The centrosome, Chap 12. New York: Springer Science and Business Media.

Krämer, A., Maier, B., and Bartek, J. (2011). Centrosome clustering and chromosomal (in)stability: A matter of life and death. Mol Oncol 5(4):324–335.

Krämer A, Anderhub S, Maier B (2012). Mechanisms and consequences of centrosome clustering in cancer cells. In: Schatten H, editor. The centrosome, Chap 17. New York: Springer Science and Business Media.

Kramer ER, Scheuringer N, Podtelejnikov AV, Mann M, Peters JM (2000). Mitotic regulation of the APC activator proteins CDC20 and CDH1. Mol Biol Cell 11:1555–1569.

Kwon M, Godinho SA, Chandhok NS, Ganem NJ, Azioune A, Thery M, Pellman D (2008). Mechanisms to suppress multipolar divisions in cancer cells with extra centrosomes. Genes Dev 22:2189–2203.

Landen Jr CN, Lin YG, Immaneni A, Deavers MT, Merritt WM, Spannuth WA, Bodurka DC, Gershenson DM, Brinkley WR, Sood AK (2007). Overexpression of the Centrosomal Protein Aurora-A Kinase is Associated with Poor Prognosis in Epithelial Ovarian Cancer Patients. Clin Cancer Res 13:4098–4104.

Lassus H, Sihto H, Leminen A, Nordling S, Joensuu H, Nupponen NN, Butzow R (2004). Genetic alterations and protein expression of KIT and PDGFRA in serous ovarian carcinoma. Br J Cancer 91:2048–2055.

Leber, B., Maier, B., Fuchs, F., Chi, J., Riffel, P., Anderhub, S., Wagner, L., Ho, A. D., Salisbury, J. L., Boutros, M., and Krämer, A. (2010). Proteins required for centrosome clustering in cancer cells. Sci Transl Med 2(33).

Lee J, Gollahon L (2013). NEK2-targeted ASO or siRNA pretreatment enhances anticancer drug sensitivity in triple-negative breast cancer cells. Int J Oncol 42:839–847. PMID: 3597451. https://doi.org/10.3892/ijo.2013.1788

Lee MY, Moreno CS, Saavedra HI (2014). E2F activators signal and maintain centrosome amplification in breast cancer cells. Mol Cell Biol 34: 2581–2599. PMID: 4097658. https://doi.org/10.1128/MCB.01688-13

Levy YY, Lai EY, Remillard SP, Heintzelman MB, Fulton C (1996). Centrin is a conserved protein that forms diverse associations with centrioles and MTOCs in Naegleria and other organisms. Cell Motil Cytoskeleton 33:298–323.

Li YW, Hussain M, Sarkar SH, Eliason J, Li R, Sarkar FH (2005). Gene expression profiling revealed novel mechanism of action of Taxotere and Furtulon in prostate cancer cells. BMC Cancer 5:7.

Li Y, Lu W, Chen D, Boohaker RJ, Zhai L, Padmalayam I, Wennerberg K, Xu B, Zhang W (2015) KIFC1 is a novel potential therapeutic target for breast cancer. Cancer Biol Ther 16:1316–1322

Li J, Diao H, Guan X, Tian X (2020). Kinesin Family Member C1 (KIFC1) Regulated by Centrosome Protein E (CENPE) Promotes Proliferation, Migration, and Epithelial-Mesenchymal Transition of Ovarian Cancer. Med Sci Monit 26:e927869. https://doi.org/10.12659/MSM.927869. PMID: 33361741; PMCID: PMC7780892.

Lin S, Zhou S, Jiang S, Liu X, Wang Y, Zheng X, Zhou H, Li X, Cai X (2016). NEK2 regulates stem-like properties and predicts poor prognosis in hepatocellular carcinoma. Oncol Rep 36:853–862. PMID: 27349376. https://doi.org/10.3892/or.2016.4896

Lingle WL, Lutz WH, Ingle JN, Maihle NJ, Salisbury JL (1998). Centrosome hypertrophy in human breast tumors: implications for genomic stability and cell polarity. Proc Natl Acad Sci USA 95:2950–2955.

Lingle WL, Salisbury JL (1999). Altered centrosome structure is associated with abnormal mitoses in human breast tumors. Am J Pathol 155:1941–1951.

Lingle WL, Salisbury JL (2000). The role of the centrosome in the development of malignant tumors. Curr Top Dev Biol 49:313–329.

Lingle WL, Barrett SL, Negron VC, D'Assoro AB, Boeneman K, Liu W, Whitehead CM, Reynolds C, Salisbury JL (2002). Centrosome amplification drives chromosomal instability in breast tumor development. Proc Natl Acad Sci USA 99:1978–1983.

Liu X, Gao Y, Lu Y, Zhang J, Li L, Yin F (2014). Up-regulation of NEK2 is associated with drug resistance in ovarian cancer. Oncol Rep 31:745–754. PMID: 24337664. https://doi.org/10.3892/or.2013.2910

Liu Y, Shen Z, Zhu T, Lu W, Fu Y (2023). Curcumin enhances the anti-cancer efficacy of paclitaxel in ovarian cancer by regulating the miR-9-5p/BRCA1 axis. Front Pharmacol 13:1014933. https://doi.org/10.3389/fphar.2022.1014933. PMID: 36703740; PMCID: PMC9871306.

Lutz W, Lingle WL, McCormick D, Greenwood TM, Salisbury JL (2001). Phosphorylation of centrin during the cell cycle and its role in centriole separation preceding centrosome duplication. J Biol Chem 276:20774–20780.

Mack GJ, Ou Y, Rattner JB (2000). Integrating centrosome structure with protein composition and function in animal cells. Microsc Res Tech 49:409–419.

Manandhar G, Schatten H, Sutovsky P (2005). Centrosome reduction during gametogenesis and its significance. Biol Reprod 72:2–13.

Marchetti F, Mailhes JB, Bairnsfather L, Nandy I, London SN (1996). Dose-response study and threshold estimation of griseofulvin induced aneuploidy during female mouse meiosis I and II. Mutagenesis 11:195–200.

Marina M, Saavedra HI (2014). NEK2 and PLK4: prognostic markers, drivers of breast tumorigenesis and drug resistance. Front Biosci 19:352–365. PMID: 24389189.

Marthiens, V., Piel, M., and Basto, R. (2012). Never tear us apart - the importance of centrosome clustering. J Cell Sci 125(14):3281–3292.

Matei D, Emerson RE, Lai YC, Baldridge LA, Rao J, Yiannoutsos C, Donner DD (2006). Autocrine activation of PDGFR alpha promotes the progression of ovarian cancer. Oncogene 25:2060–2069.

Matsumoto Y, Maller JL (2002). Calcium, calmodulin, and CaMKII requirement for initiation of centrosome duplication in Xenopus egg extracts (comment). Science 295:499–502

McNaught KS, et al (2002). Impairment of the ubiquitin-proteasome system causes dopaminergic cell death and inclusion body formation in ventral mesencephalic cultures. J Neurochem 81:301–630.

Meraldi P and Nigg EA (2001). Centrosome cohesion is regulated by a balance of kinase and phosphatase activities. J Cell Sci 114:3749–3757.

Meraldi P, Honda R, Nigg EA (2002). Aurora-A overexpression reveals tetraploidization as a major route to centrosome amplification in p53–/– cells. EMBO J 21:483–492.

Merdes A, Ramyar K, Vechio JD, Cleveland DW (1996). A complex of NuMA and cytoplasmic dynein is essential for mitotic spindle assembly. Cell 87:447–458.

Merdes A, Cleveland DA (1998). The role of NuMA in the interphase nucleus. J Cell Sci 111:71–79.

Miao YL, Zhang X, Zhao JG, Spate L, Zhao MT, Murphy CN, Prather RS, Sun QY, Schatten H (2012). Effects of griseofulvin on in vitro porcine oocyte maturation and embryo development. Environ Mol Mutagen 53(7):561–6. https://doi.org/10.1002/em.21717.

Michaud EJ, Yoder BK (2006). The primary cilium in cell signaling and cancer. Cancer Res 66:6463–6467.

Mignogna C, Staropoli N, Botta C, De Marco C, Rizzuto A, Morelli M, Di Cello A, Franco R, Camastra C, Presta I, Malara N, Salvino A, Tassone P, Tagliaferri P, Barni T, Donato G, Di Vito A (2016). Aurora Kinase A expression predicts platinum-resistance and adverse outcome in high-grade serous ovarian carcinoma patients. J Ovarian Res 9(1):31. https://doi.org/10.1186/s13048-016-0238-7. PMID: 27209210; PMCID: PMC4875597

Mittal K, Choi DH, Klimov S, Pawar S, Kaur R, Mitra AK, Gupta MV, Sams R, Cantuaria G, Rida PCG, Aneja R (2016). A centrosome clustering protein, KIFC1, predicts aggressive disease course in serous ovarian adenocarcinomas. J Ovarian Res 9:17. https://doi.org/10.1186/s13048-016-0224-0. PMID: 26992853; PMCID: PMC5477851.

Mogensen MM, Malik A, Piel M, Bouckson-Castaing V, Bornens M (2000). Microtubule minus-end anchorage at centrosomal and non-centrosomal sites: the role of ninein. J Cell Sci 113:3013–3023.

Mohamadian M, Bahrami A, Binabaj MM, Asgharzadeh F, Ferns GA (2022). Molecular targets of curcumin and its therapeutic potential for ovarian cancer, Nutrition and Cancer, 74:8, 2713-2730, https://doi.org/10.1080/01635581.2022.2049321

Montero-Conde C, Martin-Camposo JM, Lerma E, Martinez-Guitarte JL, Combalia N, Montaner D, Matias-Guiu X, Dopazo J, de Leiva A, Robledo M et al (2008). Molecular profiling related to poor prognosis in thyroid carcinoma. Combining gene expression data and biological information. Oncogene 27:1554–1561.

Moreno et al (2019). The nuclear structural protein NuMA is a negative regulator of 53BP1 in DNA double-strand break repair. Nucleic Acids Research, Volume 47, Issue 6, Pages 2703–2715, https://doi.org/10.1093/nar/gkz138

Morretton et al (2022). A catalog of numerical centrosome defects in epithelial ovarian cancers. EMBO Molecular Medicine 14: e15670, 1–12.

Münger K, Duensing S (2004). Radiation therapy and centrosome anomalities in pancreatic cancer. In: Nigg

E, editor. Centrosomes in development and disease. Weinheim: Wiley-VCA; p. 353–70.

Mukerjee A, Viswanatha JK (2009). Formulation, characterization and evaluation of curcumin-loaded PLGA nanospheres for cancer therapy. Anticancer Res 29:3867–3876.

Murphy S, Urbani L, Stearns T (1998). The mammalian gamma-tubulin complex contains homologues of yeast spindle pole body component sspc97p and spc98p. J Cell Biol 141:663–674.

Nagaraju GP, Aliya S, Zafar SF, Basha R, Diazd R, El-Rayes BF (2012). The impact of curcumin on breast cancer. Integr Biol 4(9):996–1007. https://doi.org/10.1039/c2ib20088k.

Ohki R, Nemoto J, Murasawa H, Oda E, Inazawa J, Tanaka N, Taniguchi T (2000). Reprimo, a new candidate mediator of the p53-mediated cell cycle arrest at the G2 phase. J Biol Chem 275:22627–22630.

Ohta Y, Ohba T, Miyamoto E (1990). Ca2+/calmodulin-dependent protein kinase II: localization in the interphase nucleus and the mitotic apparatus of mammalian cells. Proc Natl Acad Sci USA 87:5341–5345.

Ohta T, Essner R, Ryu JH, Palazzo RE, Uetake Y, Kuriyama R (2002). Characterization of Cep135, a novel coiled-coil centrosomal protein involved in microtubule organization in mammalian cells. J Cell Biol 156:87–99.

Olivero OA (2012). Centrosomal amplification and related abnormalities induced by nucleoside analogs. In: Schatten H, editor. The centrosome, Chap 16. New York: Springer Science and Business Media.

Ou Y, Rattner JB (2004). The centrosome in higher organisms: structure, composition and duplication. Int Rev Cytol 238:119–182.

Pan J, Snell W (2007). The primary cilium: keeper of the key to cell division. Cell 129: 1255–1257.

Panda D, Rathinasamy K, Santra MK, Wilson L (2005). Kinetic suppression of microtubule dynamic instability by griseofulvin: implications for its possible use in the treatment of cancer. Proc Natl Acad Sci USA 102:9878–9883.

Patel N, Weekes D, Drosopoulos K, Gazinska P, Noel E, Rashid M, Mirza H, Quist J, Brasó-Maristany F, Mathew S, Ferro R, Pereira AM, Prince C, Noor F, Francesch-Domenech E, Marlow R, De Rinaldis E, Grigoriadis A, Linardopoulos S, Marra P, Tutt AN (2018). Integrated genomics and functional validation identifies malignant cell specific dependencies in triple negative breast cancer. Nat Commun 9(1).

Pawar S, Donthamsetty S, Pannu V, et al (2014). KIFCI, a novel putative prognostic biomarker for ovarian adenocarcinomas: delineating protein interaction networks and signaling circuitries. J Ovarian Res 7:53

Pietromonaco SF, Seluja GA, Elias L (1995). Identification of enzymatically active Ca2+/calmodulin dependent protein kinase in centrosomes of hematopoietic cells. Blood Cells Mol Dis 21:34–41.

Pihan GA, Doxsey SJ (1999). The mitotic machinery is a source of genetic instability in cancer. Semin Cancer Biol 9:289–302.

Pockwinse SM, Krockmalnic G, Doxsey SJ, Nickerson J, Lian JB, vanWijnen AJ, Stein JL, Stein GS, Penman S (1997). Cell cycle independent interaction of CDC with the centrosome, which is associated with the nuclear matrix-intermediate filament scaffold. Proc Natl Acad Sci USA 94:3022–3027.

Prosser SL, Fry AM (2012). Regulation of the centrosome cycle by protein degradation. In: Schatten H, editor. The centrosome, Chap 9. New York: Springer Science and Business Media.

Pugacheva EN, Jablonski SA, Hartman TR, Henske EP, Golemis EA (2007). HEF1-dependent aurora A activation induces disassembly of the primary cilium. Cell 129:1351–1363.

Quarmby LM, Parker JDK (2005). Cilia and the cell cycle? J Cell Biol 169(5):707–710.

Quintyne NJ, Gill SR, Eckley DM, Crego CL, Compton DA, Schroer TA (1999). Dynactin is required for microtubule anchoring at fibroblast centrosomes. J Cell Biol 147:321–334.

Quintyne NJ, Schroer TA (2002). Distinct cell cycle-dependent roles for dynactin and dynein at centrosomes. J Cell Biol 159:245–254.

Rao Z, Zhu Y, Yang P, Chen Z, Xia Y, Qiao C, Liu W, Deng H, Li J, Ning P, Wang Z (2022). Pyroptosis in inflammatory diseases and cancer. Theranostics 12(9):4310–4329. https://doi.org/10.7150/thno.71086. PMID: 35673561; PMCID: PMC9169370.

Reader JC, Fan C, Ory EC, Ju J, Lee R, Vitolo MI, Smith P, Wu S, Ching MMN, Asiedu EB, Jewell CM, Rao GG, Fulton A, Webb TJ, Yang P, Santin AD, Huang HC, Martin SS, Roque DM (2022). Microtentacle formation in ovarian carcinoma. Cancers (Basel) 14(3):800. https://doi.org/10.3390/cancers14030800. PMID: 35159067; PMCID: PMC8834106.

Rebacz B, Larsen TO Clausen MH, Ronnest MH, Loffler H, Ho AD, Krämer A (2007). Identification of griseofulvin as an inhibitor of centrosomal clustering in a phenotype-based screen. Cancer Res 67:6342–50.

Reiter R, Gais P, Jutting U, Steuer-Vogt MK, Pickhard A, Bink K, Rauser S, Lassmann S, Höfler H, Werner M, Walch A (2006). Aurora kinase A messenger RNA overexpression is correlated with tumor progression and shortened survival in head and neck squamous cell carcinoma. Clin Cancer Res 12:5136–5141.

Reuter S, Gupta SC, Park B, Goel A, Aggarwal BB (2011). Epigenetic changes induced by curcumin and other natural compounds. Genes Nutr 6:93–108.

Rosen DG, Yang G, Liu G, Mercado-Uribe I, Chang B, Xiao XS, Zheng J, Xue FX, Liu J (2009). Ovarian cancer: pathology, biology, and disease models. Front Biosci (Landmark Ed) 14(6):2089–102. https://doi.org/10.2741/3364. PMID: 19273186; PMCID: PMC2858969.

Roth J, Yam GH, Fan J, Hirano K, Gaplovska-Kysela K, Le Fourn V, Guhl B, Santimaria R, Torossi T, Ziak M, Zuber C (2008). Protein quality control: the who's who, the where's and therapeutic escapes. Histochem Cell Biol 129:163–177.

Safavy A, Raisch PKP, Mantena S, Sanford LL, Sham SW, Krishna NR, Bonner JA (2007). Design and development of water-soluble curcumin conjugates as potential anticancer agents. J Med Chem 50:6284–6288.

Saladino C, Bourke E, Morrison CG (2012). Centrosomes, DNA damage and aneuploidy. In: Schatten H, editor. The centrosome, Chap 13. New York: Springer Science and Business Media.

Salisbury JL (1995). Centrin, centrosomes, and mitotic spindle poles. Curr Opin Cell Biol 7:39–45.

Salisbury JL, Suino KM, Busby R, Springett M (2002). Centrin-2 is required for centriole duplication in mammalian cells. Curr Biol 12:1287–1292.

Salisbury JL (2004). Centrosomes: Sfi1p and centrin unravel a structural riddle. Curr Biol 14:R27–29.

Saredi A, Howard L, Compton DA (1997). Phosphorylation regulates the assembly of NuMA in a mammalian mitotic extract. J Cell Sci 110:1287–1297.

Satir P, Christensen ST (2008). Structure and function of mammalian cilia. Histochem Cell Biol 129:687–693.

Sato N, Maitra A, Fukushima N, van Heek NT, Matsubayashi H, Iacobuzio-Donahue CA, Rosty C, Goggins M (2003). Frequent hypomethylation of multiple genes overexpressed in pancreatic ductal adenocarcinoma. Cancer Res 63:4158–66.

Sauer et al (2022). Molecular landscape and functional characterization of centrosome amplification in ovarian cancer. https://doi.org/10.1101/2022.09.13.507750

Schatten G, Schatten H, Bestor T, Balczon R (1982a). Taxol inhibits the nuclear movements during fertilization and induces asters in unfertilized sea urchin eggs. J Cell Biol 94:455–465.

Schatten H (1977). Untersuchungen über die Wirkung von Griseofulvin in Seeigeleiern und in Mammalierzellen. Universität Heidelberg; 1977 (Effects of griseofulvin on sea urchin eggs and on mammalian cells. University of Heidelberg; 1977).

Schatten H, Schatten G, Petzelt C, Mazia D (1982b). Effects of griseofulvin on fertilization and early development of sea urchins. Independence of DNA synthesis, chromosome condensation, and cytokinesis cycles from microtubule-mediated events. Eur J Cell Biol 27:74–87.

Schatten H, Wiedemeier A, Taylor M, Lubahn D, Greenberg NM, Besch-Williford C, Rosenfeld C, Day K, Ripple M (2000). Centrosomes-centriole abnormalities are markers for abnormal cell divisions and cancer in the transgenic adenocarcinoma mouse prostate (TRAMP) model. Biol Cell 92:331–340.

Schatten H (2008). The mammalian centrosome and its functional significance. Histochem Cell Biol 129:667–686.

Schatten H, Sun QY (2009). The functional significance of centrosomes in mammalian meiosis, fertilization, development, nuclear transfer, and stem cell differentiation. Environ Mol Mutagen 50(8):620–636.

Schatten H, Sun Q-Y (2010). The role of centrosomes in fertilization, cell division and establishment of asymmetry during embryo development. Semin Cell Dev Biol 21:174–184.

Schatten H, Sun Q-Y (2011) The significant role of centrosomes in stem cell division and differentiation. Microsc Microanal 17(4):506–512.

Schatten, H. (2013). The impact of centrosome abnormalities on breast cancer development and progression with a focus on targeting centrosomes for Breast Cancer Therapy. In: Cell and Molecular Biology of Breast Cancer. Edited by Heide Schatten, published by Springer Science and Business Media, LLC.

Schatten, H. (2014). The role of centrosomes in cancer stem cell functions. Chapter 12, pages 259–279. In: Cell and Molecular Biology and Imaging of Stem Cells, First Edition. Edited by Heide Schatten. © 2014 John Wiley & Sons, Inc. Published 2014 by John Wiley & Sons, Inc.

Schatten H, Sun QY. (2014). Posttranslationally modified tubulins and other cytoskeletal proteins: their role in gametogenesis, oocyte maturation, fertilization and Pre-implantation embryo development. Adv Exp Med Biol. 2014;759:57–87. https://doi.org/10.1007/978-1-4939-0817-2_(Invited book chapter: In Posttranslational Protein Modifications in the Reproductive System; Edited by Peter Sutovsky; Springer Science and Business Media)

Schatten H, Sun QY (2018). Functions and dysfunctions of the mammalian centrosome in health, disorders, disease, and aging. Histochem Cell Biol 150(4):303–325. https://doi.org/10.1007/s00418-018-1698-1. PMID: 30062583.

Schatten H (2022a). Centrosome Dysfunctions in Cancer. Adv Anat Embryol Cell Biol 235:43–50. https://doi.org/10.1007/978-3-031-20848-5_4. PMID: 36525110

Schatten H (2022b). The Centrosome Cycle within the Cell Cycle. Adv Anat Embryol Cell Biol 235:17–35. https://doi.org/10.1007/978-3-031-20848-5_2. PMID: 36525108

Scheer U (2014). Historical roots of centrosome research: discovery of Boveri's microscope slides in Würzburg. Philos. Trans. R. Soc. B 369, 20130469. https://doi.org/10.1098/rstb.2013.0469

Scheer U (2018). Boveri's research at the Zoological Station Naples: Rediscovery of his original microscope slides at the University of Würzburg. Marine Genomics https://doi.org/10.1016/j.margen.2018.01.003

Schiff PB, Fant J, Horwitz SB (1979). Promotion of microtubule assembly in vitro by taxol. Nature 277:665–667.

Schlegel BP, Starita LM, Parvin JD (2003). Overexpression of a protein fragment of RNA helicase A causes inhibition of endogenous BRCA1 function and defects in ploidy and cytokinesis in mammary epithelial cells. Oncogene 22:983–991.

Schneider L, Clement CA, Teilmann SC, et al (2005). PDGFR alpha signaling is regulated through the primary cilium in fibroblasts. Curr Biol 15:1861–1866.

Schroer TA (2001). Microtubules don and doff their caps: dynamic attachments at plus and minus ends. Curr Opin Cell Biol 13:92–96.

Shang X, Burlingame SM, Okcu MF, Ge N, Russell HV, Egler RA, David RD, Vasudevan SA, Yang J, Nuchtern JG (2009). Aurora A is a negative prognostic factor and a new therapeutic target in human neuroblastoma. Mol Cancer Ther 8:2461–2469.

Shao S, Liu R, Wang Y, et al (2010). Centrosomal Nlp is an oncogenic protein that is gene-amplified in human tumors and causes spontaneous tumorigenesis in transgenic mice. J Clin Invest 120:498–507.

Sharma N, Berbari NF, Yoder BK (2008). Ciliary dysfunction in developmental abnormalities and diseases. Curr Top Dev Biol 2008(85):371–427.

Shin SJ, Beech JR, Kelly KA (2012). Targeted nanoparticles in imaging: paving the way for personalized medicine in the battle against cancer. Integr Biol 5(1):29–42. https://doi.org/10.1039/C2IB20047C.

Shintani K, Ebisu H, Mukaiyama M, Hatanaka T, Chinen T, Takao D, Nagumo Y, Sakakura A, Hayakawa I, Usui, T (2020). Structure optimization of gatastatin for the development of γ-tubulin-specific inhibitor. ACS Med Chem Lett 11:1125–1129.

Sluder G (2004). Centrosome duplication and its regulation in the higher animal cell. In: Nigg E, editor. Centrosomes in development and disease. Weinheim: Wiley-VCA; p. 167–89.

Sorokin S (1962). Centrioles and the formation of rudimentary cilia by fibroblasts and smooth muscle cells. J Cell Biol 15:363–377.

Starita LM, Machida Y, Sankaran S, Elias JE, Griffin K, Schlegel BP, Gygi SP, Parvin JD (2004). BRCA1-dependent ubiquitination of gamma-tubulin regulates centrosome number. Mol Cell Biol 24:8457–8466.

Steadman BT, Schmidt PH, Shanks RA, Lapierre LA, Goldenring JR (2002). Transforming acidic coiled-coil-containing protein 4 interacts with centrosomal AKAP350 and the mitotic spindle apparatus. J Biol Chem 277(33):30165–30176.

Steiert B, Icardi CM, Faris R, Weber MM (2023). The Chlamydia trachomatis type III-secreted effector protein CteG induces centrosome amplification through interactions with centrin-2. PNAS Vol. 120 No. 20, e2303487120, 1–11.

Sun QY, Lai L, Park KW, Kühholzer B, Prather RS, Schatten H (2001a). Dynamic events are differently regulated by microfilaments, microtubules, and mitogen-activated protein kinase during porcine oocyte maturation and fertilization in vitro. Biol Reprod 64:879–889.

Sun QY, Lai L, Bonk A, Prather RS, Schatten H (2001b). Cytoplasmic changes in relation to nuclear maturation and early embryo developmental potential of porcine oocytes: effects of gonadotropins, cumulus cells, follicular size and protein synthesis inhibition. Mol Reprod Dev 59:192–198.

Sun Q-Y, Lai L, Wu G, Park K-W, Day B, Prather RS, Schatten H (2001c). Microtubule assembly after treatment of pig oocytes with taxol: correlation with chromosomes, γ-tubulin and MAP kinase. Mol Reprod Dev 60:481–490.

Sun Q-Y, Wu GM, Lai L, Park KW, Day B, Prather RS, Schatten H (2001d). Translocation of active mitochondria during porcine oocyte maturation, fertilization and early embryo development in vitro. Reproduction 122:155–163.

Sun Q-Y, Lai L, Wu G, Bonk A, Cabot R, Park K-W, Day B, Prather RS, Schatten H (2002). Regulation of mitogen-activated protein kinase phosphorylation, microtubule organization, chromatin behavior, and cell cycle progression are regulated by protein phosphatases during pig oocyte maturation and fertilization in vitro. Biol Reprod 66(3):580–588.

Sun QY, Schatten H (2006). Multiple roles of NuMA in vertebrate cells: Review of an intriguing multifunctional protein. Front Biosci 11:1137–1146.

Sun Q-Y, Schatten H (2007). Centrosome inheritance after fertilization and nuclear transfer in mammals. In: Sutovsky P (ed) Somatic cell nuclear transfer. Landes Bioscience. Adv Exp Med Biol 591:58–71.

Syed MI, Syed S, Minty F, Harrower S, Singh J, Chin A, McLellan DR, Parkinson EK, Clark LJ (2009). Gamma tubulin: A promising indicator of recurrence in squamous cell carcinoma of the larynx. Otolaryngol Head Neck Surg 140:498–504.

Takahashi M, Yamagiwa A, Nishimura T, Mukai H, Ono Y (2002). Centrosomal proteins CG-NAP and kendrin provide microtubule nucleation sites by anchoring gamma-tubulin ring complex. Mol Biol Cell 13:3235–3245.

Tanner MM, Grenman S, Koul A, Johannsson O, Meltzer P, Pejovic T, Borg A, Isola JJ (2000). Frequent amplification of chromosomal region 20q12-q13 in ovarian cancer. Clin Cancer Res 6:1833–1839.

Tassin AM, Celati C, Moudjou M, Bornens M (1998). Characterization of the human homologue of the yeast Spc98p and its association with gamma-tubulin. J Cell Biol 141:689–701.

Teixidó-Travesa N, Roig J, Lüders J (2012). The where, when and how of microtubule nucleation—one ring to rule them all. J Cell Sci 125:4445–4456.

Tong C, Fan H-Y, Lian L, Li S-W, Chen D-Y, Schatten H, Sun Q-Y (2002). Polo-like kinase-1 is a pivotal regulator of microtubule assembly during mouse oocyte meiotic maturation, fertiliza tion, and early embryonic mitosis. Biol Reprod 67:546–554.

Tong C, Fan H-Y, Li S-W, Chen D-Y, Song X-F, Schatten H, Sun Q-Y (2003). Effects of MEK inhibitor U0126 on meiotic progression in mouse oocytes: microtubule organization, asymmetric division and metaphase II arrest. Cell Res 13(5):375–383.

Tonnesen HH, Masson M, Loftsson T (2002). Studies of curcumin and curcuminoids. XXVII. Cyclodextrin complexation: solubility, chemical and photochemical stability. Int J Pharm 244:127–135.

Tugendreich S, Tomkiel J, Earnshaw W, Hieter P (1995). CDC27Hs colocalizes with CDC16Hs to the centrosome and mitotic spindle and is essential for the metaphase to anaphase transition. Cell 81:261–268.

Tymon-Rosario J, Adjei NN, Roque DM, Santin AD (2021). Microtubule-interfering drugs: current and future roles in epithelial ovarian cancer treatment. Cancers (Basel) 13(24):6239. https://doi.org/10.3390/cancers13246239. PMID: 34944858; PMCID: PMC8699494.

Uen YH, Liu DZ, Weng MS, Ho YS, Lin SY (2007). NF-kappa B pathway is involved in griseofulvin-

induced G2/M arrest and apoptosis in HL-60 cells. J Cell Biochem 101(5):1165–1175.
Veland IR, Awan A, Pedersen LB, Yoder BK, Christensen ST (2009). Primary cilia and signaling pathways in mammalian development, health and disease. Nephron Physiol 111:39–53.
Wakefield J, Huang J-Y, Raff JW (2000). Centrosomes have a role in regulating the destruction of cyclin B in early Drosophila embryos. Curr Biol 10:1367–1370.
Wang Q, Xie S, Chen J, Fukasawa K, Naik U, Traganos F, Darzynkiewicz Z, Jhanwar-Uniyal M, Dai W (2002). Cell cycle arrest and apoptosis by human polo-like kinase 3 is mediated through perturbation of microtubule integrity. Mol Cell Biol 22(10):3450–3459.
Wang X, Zhou YX, Qiao W, et al (2006). Overexpression of aurora kinase A in mouse mammary epithelium induces genetic instability preceding mammary tumor formation. Oncogene 25:7148–7158.
Wang D, Veena MS, Stevenson K, Tang C, Ho B, Suh JD, Duarte V, Faull KF, Mehta K, Srivatsan ES, Wang MB (2008). Liposome-encapsulated curcumin suppresses growth of head and neck squamous cell carcinoma in vitro and in xenografts through the inhibition of nuclear factor kappa B by an AKT-independent pathway. Clin Cancer Res 14:6228–6236.
Wang HF, Takenaka K, Nakanishi A, et al (2011). BRCA2 and nucleophosmin coregulate centro some amplification and form a complex with the Rho effector kinase ROCK2. Cancer Res 71:68–77.
Wehland J, Herzog W, Weber K (1977). Interaction of griseofulvin with microtubules, microtubule protein and tubulin. J Mol Biol 111:329–342.
Wei Y, Multi S, Yang CR, Ma J, Zhang QH, Wang ZB, Li M, Wei L, Ge ZJ, Zhang CH, Ouyang YC, Hou Y, Schatten H, Sun QY (2011). Spindle assembly checkpoint regulates mitotic cell cycle progression during preimplantation embryo development. PLoS One 6(6):e21557
Wen S, Liu Y, Yang M, Yang K, Huang J, Feng D (2016). Increased NEK2 in hepatocellular carcinoma promotes cancer progression and drug resistance by promoting PP1/Akt and Wnt activation. Oncol Rep 36:2193–2199. PMID: 27509921. https://doi.org/10.3892/or.2016.5009
Wheatley DN, Wang AM, Strugnell GE (1996). Expression of primary cilia in mammalian cells. Cell Biol Int 20:73–81.
Wilken R, Veena MS, Wang MB, Srivatsan ES (2011). Curcumin: a review of anti-cancer properties and therapeutic activity in head and neck squamous cell carcinoma. Mol Cancer 10(12):1–19.
Wilkinson CJ, Andersen JS, Mann M, Nigg EA (2004). A proteomic approach to the inventory of the human centrosome. In: Nigg E, editor. Centrosomes in development and disease. Weinheim: Wiley-VCA; p. 125–42.
Woodruff JB, Wueseke O, Hyman AA (2014). Pericentriolar material structure and dynamics. Philos Trans R Soc Lond B Biol Sci 369, 20130459.
Wojcik C, DeMartino GN (2003). Intracellular localization of proteasomes. Int J Biochem Cell Biol 35:579–589.
Xia J, Franqui Machin R, Gu Z, Zhan F (2015). Role of NEK2A in human cancer and its therapeutic potentials. Biomed Res Int:862461. PMID: 4330945. https://doi.org/10.1155/2015/862461
Xu X, Weaver Z, Linke SP, Li C, Gotay J, Wang XW, Harris CC, Ried T, Deng CX (1999). Centrosome amplification and a defective G2-M cell cycle checkpoint induce genetic insta bility in BRCA1 exon 11 isoform-deficient cells. Mol Cell 3:389–395.
Yallapu MM, Gupta BK, Jaggi M, Chauhan SC (2010a). Fabrication of curcumin encapsulated PLGA nanoparticles for improved therapeutic effects in metastatic cancer cells. J Colloid Interface Sci 351(1):19–29.
Yallapu MM, Jaggi M, Chauhan SC (2010b). Poly(β-cyclodextrin)/curcumin self-assembly: a novel approach to improve curcumin delivery and its therapeutic efficacy in prostate cancer cells. Macromol Biosci 10:1141–1151.
Yam HF, Wang ZH, Or PC, Wang SW, Li J, Chew EC(1998). Effect of glucocorticoid hormone on nuclear matrix in cervical cancer cells in vitro. Anticancer Res 18:209–216.
Yamaguchi M, Moore TW, Sun A, Snyder JP, Shoji M (2012). Novel curcumin analogue UBS109 potently stimulates osteoblastogenesis and suppresses osteoclastogenesis through Smad activation and NFκB inhibition. Integr Biol 4(8):905–913.
Yan B, Chng W-J (2012). The role of centrosomes in multiple myeloma. In: Schatten H, editor. The centrosome, Chap 15. New York: Springer Science and Business Media.
Yang Y, Zhou W, Xia J, Gu Z, Wendlandt E, Zhan X, Janz S, Tricot G, Zhan F (2014). NEK2 mediates ALDH1A1-dependent drug resistance in multiple myeloma. Oncotarget 5:11986–11997. PMID: 4322982.
Yin S, Bhattacharya R, Cabral F (2010). Human mutations that confer paclitaxel resistance. Mol Cancer Ther 9:327–335.
Yoon H, Liu RH (2007). Effect of selected phytochemicals and apple extracts on NF–kB activation in human breast cancer MCF-7 cells. J Agric Food Chem 55(8):3167–3173.
Young A, Dictenberg JB, Purohit A, Tuft R, Doxsey SJ (2000). Cytoplasmic dynein-mediated assembly of pericentrin and γ tubulin onto centrosomes. Mol Biol Cell 11:2047–2056.
Zhou W, Yang Y, Xia J, Wang H, Salama ME, Xiong W, Xu H, Shetty S, Chen T, Zeng Z, Shi L, Zangari M, Miles R, Bearss D, Tricot G, Zhan F (2013). NEK2 induces drug resistance mainly through activation of efflux drug pumps and is associated with poor prognosis in myeloma and other cancers. Cancer Cell 23:48–62. PMID: 23328480. https://doi.org/10.1016/j.ccr.2012.12.001

# Drug-Resistant Epithelial Ovarian Cancer: Current and Future Perspectives

4

Megha Mehrotra, Pratham Phadte, Priti Shenoy, Sourav Chakraborty, Sudeep Gupta, and Pritha Ray

**Abstract**

Epithelial ovarian cancer (EOC) is a complex disease with diverse histological subtypes, which, based on the aggressiveness and course of disease progression, have recently been broadly grouped into type I (low-grade serous, endometrioid, clear cell, and mucinous) and type II (high-grade serous, high-grade endometrioid, and undifferentiated carcinomas) categories. Despite substantial differences in pathogenesis, genetics, prognosis, and treatment response, clinical diagnosis and management of EOC remain similar across the subtypes. Debulking surgery combined with platinum-taxol-based chemotherapy serves as the initial treatment for High Grade Serous Ovarian Carcinoma (HGSOC), the most prevalent one, and for other subtypes, but most patients exhibit intrinsic or acquired resistance and recur in short duration. Targeted therapies, such as anti-angiogenics (e.g., bevacizumab) and PARP inhibitors (for BRCA-mutated cancers), offer some success, but therapy resistance, through various mechanisms, poses a significant challenge. This comprehensive chapter delves into emerging strategies to address these challenges, highlighting factors like aberrant miRNAs, metabolism, apoptosis evasion, cancer stem cells, and autophagy, which play pivotal roles in mediating resistance and disease relapse in EOC. Beyond standard treatments, the focus of this study extends to alternate targeted agents, including immunotherapies like checkpoint inhibitors, CAR T cells, and vaccines, as well as inhibitors targeting key oncogenic pathways in EOC. Additionally, this chapter covers disease classification, diagnosis, resistance pathways, standard treatments, and clinical data on various emerging approaches, and advocates for a nuanced and personalized approach tailored to individual subtypes and resistance mechanisms, aiming to enhance therapeutic outcomes across the spectrum of EOC subtypes.

**Keywords**

Ovarian cancer · Drug resistance · mRNAs · Metabolism · Apoptosis evasion · Cancer stem cells · Autophagy · Disease relapse · Immunotherapies · Checkpoint inhibitors ·

M. Mehrotra · P. Phadte · P. Shenoy · S. Chakraborty
P. Ray (✉)
Imaging Cell Signalling & Therapeutics Lab, Advanced Centre for Treatment, Research and Education in Cancer-Tata Memorial Centre, Navi Mumbai, India

Homi Bhabha National Institute, Mumbai, India
e-mail: pray@actrec.gov.in

S. Gupta
Homi Bhabha National Institute, Mumbai, India

Department of Medical Oncology, Tata Memorial Centre, Mumbai, India



H. Schatten (ed.), *Cell and Molecular Biology of Ovarian Cancer*, Advances in Experimental Medicine and Biology 1452, https://doi.org/10.1007/978-3-031-58311-7_4

Vaccines · Oncogenic pathways · Disease classification · Resistance pathways · Personalized cancer therapies

## 4.1 Introduction to Epithelial Ovarian Cancer and Its Subtypes

Ovarian cancer (OC) is one of the deadliest gynecological malignancies, accounting for ≃2,00,000 deaths annually worldwide (Sung et al. 2021). Many studies suggest the absence of precursor lesions in the ovaries, leading to the belief that ovarian cancer develops "de novo" (Bell and Scully 1994). This highlights the lacunae in understanding the early stage of carcinogenesis in ovarian cancer. Over the years, multiple attempts at early detection and designing novel therapeutics were made, which remained unsuccessful due to poor understanding and complexity of the disease. In this chapter, we discuss the current understanding of the type, origin, diagnosis, therapeutic approaches, resistance mechanisms involved, and the upcoming drugs in clinical trials for the treatment of ovarian cancer.

Approximately ~98% of ovarian tumour develop from the thin epithelial lining of the ovary, which comprises a diverse collection of histologically and genetically distinct diseases classified into core five subtypes (Shih et al. 2021). Though the histopathological classifications of Epithelial Ovarian Cancer (EOC) are well recognized and still used in the clinic, it does not explain the extreme heterogeneity (both morphologically and genetically) observed in different subtypes, the simultaneous presence of highly aggressive, genetically instable, advanced metastatic disease with indolent, genetically stable disease, confined to the ovarian site under the same histological group and absence of any precursor lesions (Shih et al. 2021). Recent studies at the molecular level and disease pathogenesis have led to a significantly improved understanding of these subtypes of EOC thereby opening up the scope for new treatment, screening, and prevention to reduce the mortality due to this devastating disease (Kurman and Shih 2016). In 2004, Kurman and Shih proposed a new classification of EOC that focused more on the genetic aberrations and pathogenesis of the disease than the pathological characteristics and grouped EOC into two broad categories Type I and Type II. This dualistic model of EOC was officially accepted by WHO in 2014 which also accounts for the extra ovarian origin of epithelial ovarian cancer.

As per this classification, EOC can be divided into two major subtypes, i.e., Type I and Type II (Shih Ie and Kurman 2004).

### 4.1.1 Type I Neoplasms

Type I EOC progresses step-wise from premalignant to malignant lesions and usually consists of low-grade, less aggressive tumours detected at an early stage. They are genetically stable and harbor a low incidence of *TP53* mutations and frequent somatic alterations in *KRAS, PTEN, PIK3CA, MAP, ERK, BRAF, ARID1A*, and *CTNNB1* genes. Low-grade serous, endometrioid, mucinous, sero-mucinous, undifferentiated, clear cell, and malignant Brenner tumours fall under the type I category (Salazar et al. 2018).

#### 4.1.1.1 Low-Grade Serous Ovarian Cancer (LGSOC) Subtype

LGSOC subtype is relatively uncommon and accounts for no more than 15% of cases of serous carcinomas. They are diagnosed at a younger age and are primarily unresponsive to standard chemotherapeutics. LGSOCs are thought to arise in a stepwise manner from serous borderline tumours and could be either invasive or noninvasive in nature. The noninvasive tumours are usually bilateral, along with peritoneal implants in approximately 30% of the cases. However, the mechanism underlying the development of implants is elusive. Due to the presence of identical *BRAF* or *KRAS* mutations as that of the primary tumour in the implants, they may indeed be a metastatic extension of the primary ovarian disease. LGSOC majorly harbors wild-type p53, is genomically stable, and shows fewer gene copy number alterations. In more than half of the cases, activating somatic mutations in the *KRAS*-

*BRAF-MAPK* pathway are present, suggesting that LGSOC patients may benefit from MAPK-targeted agents (Soslow 2008). Mutations in *BRAF, KRAS*, and *ERBB2* are mutually exclusive and occur early in the development of LGSOC, thereby supporting the hypothesis that these are critical mutations in the progression of LGSOC (Gadducci and Cosio 2020).

#### 4.1.1.2 Sero-mucinous Subtype

They are also known as mixed Mullerian carcinomas and are generally noninvasive. The immunohistochemical profile shows the presence of ER (100%), CA-125 (92%), CK7 (100%), and PR (67%) (Taylor and McCluggage 2015). Frequent mutations in the *ARID1A* gene are also observed.

#### 4.1.1.3 Mucinous Ovarian Carcinomas (MOC)

Mucinous ovarian tumours, as the name suggests, show the presence of cysts and glands surrounded by epithelial cells containing intra-cytoplasmic mucin. MOC tumours are known to be associated with endometriosis and account for about 3% of EOC cases. This extremely uncommon cancer type is intrinsically resistant and shows poor response to the standard chemotherapy regimen of platinum-taxane, but due to early detection, it presents with an excellent prognosis (Hess et al. 2004). Metastatic MOCs have a poorer outcome. However, both primary and metastatic MOC are managed by surgery. The most common molecular alterations in this subtype include mutations in the MAP kinase pathway (49% in *KRAS* and 3.5% in *BRAF*). Furthermore, *ERBB2* amplifications (~20%), *PIK3CA/PTEN* mutations (~9%), and strong expression of ER in 10–15% and Paired Box 8 (*PAX8*) have been observed in approximately 50% of the cases, indicating targeted therapy could be promising for MOC.

#### 4.1.1.4 Endometrioid Ovarian Carcinomas (EEOC)

Endometrioid ovarian tumours, closely resembling the endometrium, constitute ~10% of all EOC. EEOC tumours remain localized to the ovary and are also associated with endometriosis. In most cases, endometrioid carcinomas are well differentiated. However, a small subset of patients also shows the presence of moderately or poorly differentiated endometrioid carcinomas. It is commonly observed that well-differentiated areas are present between moderately and poorly differentiated tumours, indicating dedifferentiation from low-grade carcinomas. EEOC is frequently seen in patients with Lynch syndrome. Somatic mutations in *CTNNB1, PTEN*, and *PIK3CA* are also observed, which show the involvement of Akt and Wnt signaling in EEOC. In approximately 60% of the cases, the mutations in *CTNNB1* affect serine or threonine residues, which are phosphorylation targets for glycogen synthase kinase 3-beta (*GSK3β*) (McConechy et al. 2014). Microsatellite instability is also observed in approximately 20% of endometrioid carcinoma patients, leading to the loss of specific essential genes like *hMLH1, hMSH2, MSH6,* and *PMS2* (Hollis et al. 2020). Mutations in *ARID1A* gene, mainly involved in the formation of the SWI/SNF chromatin complex, may lead to the transformation of endometriosis into cancer. Almost 85–95% of EEOC are of grades I and II and are thereby manageable with primary debulking surgery and have better survival outcome (Chen et al. 2021).

#### 4.1.1.5 Clear Cell Ovarian Carcinomas (CCOC)

Ovarian clear cell tumours comprise polygonal cells with distinct cell membrane and abundant lipids and glycogen in the cytoplasm. CCOC carcinoma is a rare subtype and accounts for <10% of EOC. These solid tumours often arise in an endometriotic cyst that contains chocolate-colored fluid, and the color of the neoplastic tissue may range from white and yellow to light brown. In clear cell carcinomas, the ovary is the primary site of the tumours, and an extra ovarian extension may have an aggressive clinical outcome. Unlike other type I tumours, clear cell carcinoma is not graded and is generally considered high grade. They have a very distinct methylation profile as compared to other subtypes. Activating mutations in *PIK3CA* suggest that PI3K-AKT-mTOR may be a therapeutically viable target for clear cell carcinoma (Gadducci et al. 2021).

However, the most reported aberration found in CCOC is *ARID1A* homozygous truncating mutation, which occurs in almost 57% of tumours, according to whole genome sequencing study (Jones et al. 2010). Loss of promoter methylation in the genes involved in the *HNF1* pathway is also a characteristic of CCOC, as detected by immunohistochemistry.

#### 4.1.1.6 Brenner Tumours

Most of the Brenner tumours are benign and rarely malignant. The molecular profiling of malignant tumours has not been performed extensively. Studies have reported the loss of p16 in all atypical proliferative tumours but not in benign ones, suggesting a role of p16 in transforming benign tumours into malignant ones (Kuhn et al. 2014).

Clinical presentation and treatment sensitivity for type I neoplasms: They usually present as large, unilateral neoplasms and are cystic in nature. The prognosis is excellent when the disease is confined to the ovaries. However, in advanced stages, type I tumours show an abysmal prognosis, as most tumours are refractory and show poor response to therapy.

Clinically, it is observed that women with type I EOC are younger, asymptomatic, and have lower CA-125 levels than women with type II EOC (Alcázar et al. 2013). They account for approximately 10% of deaths caused by ovarian cancer.

### 4.1.2 Type II Neoplasms

The origin of type II tumours is debatable, and some theories suggest they are extra ovarian in origin, arising from the fallopian tube and the peritoneum, and these precursor lesions include serous tubal intraepithelial carcinoma (STIC). The volume of tumours in the ovary is lesser than that of type I tumours; however, the extra ovarian disease is much greater in type II neoplasms (Koshiyama et al. 2014).

Type II tumours are high-grade in nature and are subdivided into two types, i.e., the usual and the SET (solid, pseudo- endometrioid, and transitional). Type II neoplasms show high chromosomal instability, extensive p53 mutations, and copy number amplifications in DNA repair and Notch pathways. In type II neoplasms, somatic alterations in *RB1, FOXM1, and CCNE1* are also frequent. Women with hereditary BRCA1/2 mutations usually have type II tumours (Mathieu et al. 2018).

#### 4.1.2.1 High-Grade Serous Ovarian Cancer (HGSOC)

HGSOC subtype accounts for nearly 85–90% of serous carcinomas and almost 70% of EOC cases. They are generally detected at an advanced stage, which also contributes to the overall poor prognosis. Though classically termed ovarian cancer, most HGSOC is now believed to originate from the distal fallopian tube, which migrates to the ovary due to proximity at a very early stage. This subtype shows chromosomal instability with frequent mutations in tumour suppressor p53 (>80% of the cases), disruption in the DNA repair pathway, and a high frequency of somatic copy number variations due to deletions in *RB1*, *CDKN2A/B*, and *PTEN* genes (Emmanuel et al. 2014). Alterations in the form of copy number amplifications in multiple members of *PI3KCA/AKT*, *NOTCH3*, and *FOXM1* signaling are also found to be involved in HGSOC pathophysiology. Based upon gene expression analysis of >300 HGSOC cases, this subtype is further classified into four molecular subtypes: Immunoreactive, Differentiated, Proliferative, and Mesenchymal. Each of these subtypes is associated with distinct clinical outcomes (Bell et al. 2011).

#### 4.1.2.2 High Grade EEOC

The incidence rate of high grade EEOC or grade 3 endometrioid carcinomas is rare, representing approximately 5–19% of the total endometrioid ovarian carcinomas. They are usually diagnosed at a late stage and show peritoneal carcinomatosis, ascites formation, and the presence of pelvic mass (Matulonis et al. 2016). The molecular profile of high grade EEOC is very similar to that of HGSOC. High grade EEOC show a deficiency in homologous recombination pathway and is asso-

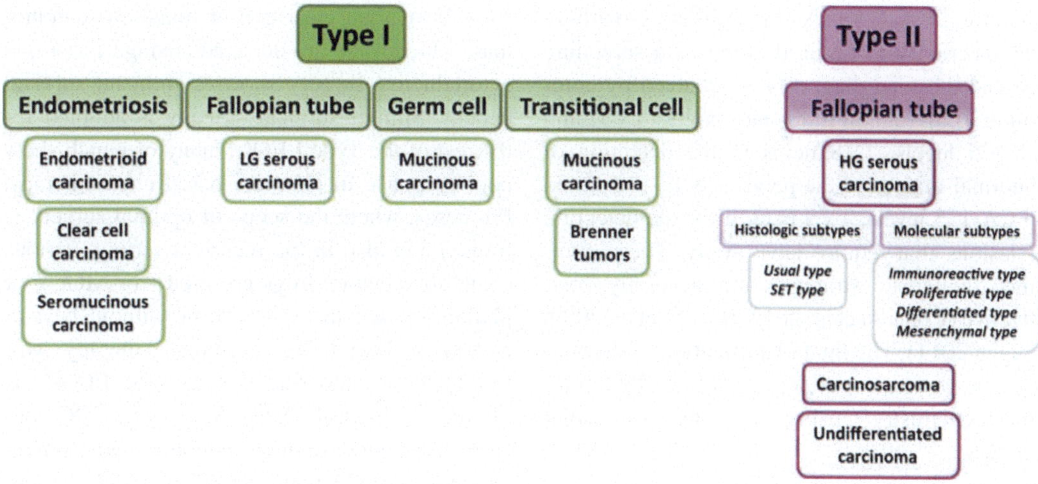


**Fig. 4.1** An expanded dualistic model of ovarian carcinogenesis. (Adopted from Kurman and Shih 2016)

ciated with *BRCA1/2* and *TP53* mutations. Histopathological analysis of high grade EEOC has revealed the presence of peripheral palisading of solid islands, along with squamous metaplasia and atypical proliferative endometrioid tumours. However, on the basis of histomorphology alone, the diagnosis of high grade EEOC can be challenging, and therefore, it is generally done in conjunction with immunohistochemical markers such as *WT1* and *TP53* to distinguish it from HGSOC (Hollis et al. 2020; Vang et al. 2009). High grade EEOC patients show a poor clinical outcome as compared to their low grade counterparts (Assem et al. 2018).

#### 4.1.2.3 Undifferentiated Carcinoma

These are a very uncommon group of tumours that do not resemble high-grade endometroid carcinomas or HGSOC. Molecular studies delineating the genetic profile of these tumours are currently unavailable (Nasioudis et al. 2017).

Clinical presentation and treatment sensitivity for type II neoplasms: More than 75% of type II neoplasms are usually detected at advanced stage. They are bilateral (both ovaries involved), rapidly proliferating, aggressive, and often accompanied by ascites formation, which is rare in type I tumours. Chemotherapy and surgery usually lead to the lengthening of progression-free survival and, in some cases, the overall survival of women diagnosed with type II tumours, but eventually, all of them succumb to this deadly disease. They account for approximately 90% of the deaths caused by ovarian cancer. The primary response to chemotherapy is good, but most tumours recur within 1–2 years (Kurman and Shih 2016) (Fig. 4.1).

## 4.2 Current Diagnostic Strategies

Despite a complete change in EOC classification in recent time, the screening or diagnostic strategies have remained unaltered over decades in clinic. EOC is initially diagnosed by increased CA-125 level in blood. Cancer-associated antigen-125 (CA-125) is secreted by coelomic epithelium and is regularly overexpressed in 90% of advanced stage ovarian carcinomas and in 50% of early ovarian cancers. However, processes that disrupt the peritoneum's epithelial lining could also lead to an elevation in CA-125 levels. Therefore, a rise in CA-125 levels could also be due to several general conditions like inflammation, pregnancy, appendicitis, infection, and cancers of the colon, lung, and pancreas (Omar and Rager 2003). CA-125 shows better efficiency in detecting Type-II EOC than Type-I and specifically low efficiency for the mucinous subtype.

Therefore, CA-125 lacks the absolute sensitivity and specificity to be used alone as a screening test and the final diagnosis is achieved by combining Transvaginal ultrasonography (TVS) and CA-125 levels. TVS helps in the detection of abnormal masses in the pelvic cavity, and based on CA-125 levels, TVS is used for the detection of lesions that require laparotomy. These combined diagnostic strategies are generally cost-effective (Chandra et al. 2019; Bast Jr. et al. 2009; Cragun 2011). The final identification of the subtypes, however, is routinely performed by immunohistochemistry using various molecular markers.

## 4.3 Current Treatment Modalities

Surgery and chemotherapy are the front-line treatments applied to all subtypes of EOC. In the majority of patients with HGSOC (the most common subtype of EOC), the disease presentation is usually at an advanced stage and involves extensive peritoneal metastasis. Treatment involves neo adjuvant chemotherapy, where the patient first receives a few cycles of chemotherapy to reduce the tumour burden followed by debulking surgery and post-operative chemotherapy to take care of any residual disease. This line of treatment reduces the disease burden, ascites generation, and allows for precise and efficient surgical removal of the tumours. Although the different histologic subtypes are distinct in their aetiology, molecular landscape, and prognosis, they are often treated as a single entity in clinics. The standard first-line therapy regimen comprises cytoreductive surgery combined with platinum-paclitaxel-based chemotherapy. However, there is a distinct difference in the platinum response between Type I and Type II EOCs. While most of Type I subtypes are inherently resistant to platinum at various degrees, the Type II HGSOC patients show an excellent response of 70–80% at the initial stage but subsequently develop resistance and relapse (Ozols 2006).

In case of other subtypes of EOC, the extent of treatment is dependent upon the stage of the disease. In ovarian clear cell or mucinous carcinomas, which often are detected at stage I, optimal cytoreductive surgery shows significant survival benefit. Similar surgical strategy is adopted for the rest of the Type I EOC, many of which show poor response to platinum-based chemotherapy. For cases, where the scope of optimal surgery is limited and also in the recurrent setting, various chemotherapeutic drugs are used (for MOC), or platinum-based therapy with or without bevacizumab followed by hormonal therapy (for LGSOC) or endocrine therapy (for EEOC) is offered (Lisio et al. 2019). Surgery in EOC consists of bilateral salpingo-oophorectomy, omentectomy, lymph node sampling, and tumour debulking (Agarwal and Kaye 2003; Ozols 2000).

In recurrent HGSOC patients, the choice of chemotherapy is dependent upon the Platinum Free Interval (PFI), which is defined as the duration between the last platinum treatment and the detection of relapse. Based on the duration of the PFI, patients are stratified as platinum-refractory (disease recurrence during therapy or within 4 weeks of last treatment cycle), platinum-resistant (disease recurring within 6 months of last treatment cycle), partially platinum-sensitive (disease recurring between 6 and 12 months of last treatment cycle), platinum-sensitive (disease progressing after 12 months of last treatment cycle) (Luvero et al. 2014). At present, for platinum-sensitive patients, if the relapse occurs after 6–24 months of first-line therapy, the patients are re-challenged with platinum-based agents. However, in platinum-resistant and refractory cases, that is, those patients who relapse sooner than 6 months, sequential use of a single agent therapy with a non-platinum compound is carried out. The non-platinum agents consist of drugs interfering with DNA replication or DNA synthesis or microtubule formation and are majorly used for second or third-line therapy for EOC. The four commonly used drugs gemcitabine, pegylated liposomal doxorubicin, paclitaxel, and topotecan show similar response rates, progression-free survival (PFS) and overall survival (OS) (Clamp and Jayson 2002). However, the choice of an individual agent prescribed by the clinician depends upon the history of the patient, cost, toxicity, availability, and convenience of treatment (Pujade-Lauraine et al. 2019).

### 4.3.1 Targeted Therapy

Lack of targeted therapy is a major constraint for successful treatment of EOC. Though recent whole genome analyses discovered many genetic aberrations for different subtypes that might act as potential therapeutic targets, translation of such knowledge to the clinic is still awaiting. To date, HGSOC is the only subtype where two distinct targeted therapies are routinely offered. Bevacizumab, a monoclonal antibody against VEGF-A is used as a maintenance therapy in combination with paclitaxel, pegylated liposomal doxorubicin (PLD), or topotecan for both platinum-resistant and platinum-sensitive HGSOC patients. The AURELIA study has demonstrated the benefit of bevacizumab in the platinum-resistant setting. Monotherapy of bevacizumab has also shown some clinical benefit in platinum-resistant patients with a response rate of 16–21% (Gupta et al. 2019).

The other therapeutic drugs that exclusively target the *BRCA1/2* mutant population are the first and second generations of PARP-1 (poly(ADP)-ribose polymerase) inhibitors (Willmott and Fruehauf 2010). Olaparib, rucaparib, and niraparib are PARP inhibitors FDA-approved for the treatment of EOC. PARP plays a critical role in NER or BER pathways and allows cells to repair the DNA damage caused by alkylating agents and other drugs. Inhibition of PARP leads to stalling of replication forks, resulting in highly cytotoxic Double Stranded Breaks (DSBs). Tumour cells of HGSOC patients harboring mutations in *BRCA1* and *BRCA2* have deficient Homologous recombination (HR) and thereby exclusively depend upon NER and BER pathways when treated with platinum agents. Treatment with Olaparib or other PARP-1 inhibitors in HRD background exerts a synthetic lethal condition leading to efficacious tumour cell death. The principle, clinical significance, and future of PARP-1 inhibitors have been discussed in detail in later sections.

Another drug, Mirvetuximab soravtansine (MIRV), has recently received (2022) FDA approval for the treatment of platinum-resistant ovarian cancer with high expression of Folate receptor (FR) (Birrer et al. 2019). This FR-α targeting MIRV is an antibody drug conjugate that delivers the cytotoxic drug, DM4 (soravtansine) after internalization of the antibody. DM4 binds to tubulin and leads to disruption of the microtubule assembly dynamics, ultimately leading to cell cycle arrest and apoptosis (Wang et al. 2022; Ponte et al. 2016).

## 4.4 Resistance, a Major Limitation Against Currently Available Therapeutics

Most women suffering from EOC develop resistance to disease, which recur despite initially responding to treatment. Resistance to drugs is the major cause of treatment failure and death in over 90% of the patients with metastatic disease. Additionally, each of the EOC subtypes shows differential chemosensitivity with distinct mechanisms of treatment resistance. Clear cell, mucinous, endometroid, and LGSOC tumours have low response rates to standard treatment modalities, and a significant proportion of patients show a high degree of primary resistance. In contrast, women suffering from HGSOC despite showing good initial response to treatment eventually recur and develop chemoresistance. Thus, drug resistance patterns in EOC are complex (Marchetti et al. 2021).

The susceptibility of cancer cells to chemotherapeutic agents like cisplatin and carboplatin is determined by the ability of cancer cells to detect DNA damage and repair it. Chemotherapeutics generally target rapidly proliferating cells, which form the bulk of the tumour population. Initially, most of the neoplastic cells in the tumour appear to respond to treatment. However, the residual population enables the growth of tumours that no longer respond to chemotherapeutic drugs (Cree and Charlton 2017). Primary or intrinsically resistant patients generally possess HR proficient tumours.

Additionally, accumulation of somatic mutations and epigenetic changes with time confer resistance to the chemotherapeutic agents, generally termed acquired resistance (Agarwal and Kaye 2003). EOC represents an extremely heterogenous group of tumours and within each tumour of a specific subtype, intratumoural heterogeneity (ITH) contributes to therapy resistance. Each tumour shows heterogenous patterns of immunostaining, and not every mutation is present in every cancer cell within the bulk of the tumour (Kossaï et al. 2018). Therefore, it becomes important to study the underlying drug resistant mechanisms, which will be highlighted in the next section of the chapter.

## 4.5 Elucidating Drug Resistance Mechanisms: Understanding the Molecular Players

Most patients who respond initially to platinum treatment become resistant over time, and this acquired resistance is due to different molecular mechanisms. Some of the molecular mechanisms are briefly summarized below:

### 4.5.1 Decreased Intracellular Drug Accumulation and Increased Drug Export

Reducing cellular accumulation by avoiding drug exposure is cancer cells' first defense mechanism. This reduced accumulation of drugs inside the cells might be due to decreased influx or increased efflux caused by the activity of various membrane transporters.

#### 4.5.1.1 Influx Transporters

These transporters mainly belong to the Solute-Like-Carrier (SLC) family and play an essential role in increasing the intracellular accumulation of drugs. SLC22 and SLC31 subfamilies are the cell's most prominent accumulators of platinum derivatives (Elsnerova et al. 2016). The cellular expression of SLC22A5 was downregulated in tumours of EOC. Lower SLC22A5 expression was observed in patients with advanced HGSOC ovarian carcinoma, indicating a potentially harmful role in disease progression (Larson et al. 2009). Inhibition of SLC31A1 decreased intracellular drug levels and increased chemoresistance in cell lines and mouse models (Holzer et al. 2004). Overexpression of SLC31A1 in A2780, an ovarian cancer cell line, led to a 55% increase in cisplatin accumulation after 24 h (Liang et al. 2012). Elevated levels of SLC31A1 after chemotherapy were positively associated with better clinical outcomes (Szakács et al. 2006). SLC31A1 is positively co-related with IGF1R and prognosticated prolonged survival in high grade serous ovarian cancer patients (Sun et al. 2016).

#### 4.5.1.2 Efflux Transporters

ATP-binding cassette (ABC) transporters facilitate the efflux of various anticancer drugs and are associated with the acquirement of chemoresistance in several cancers. Their levels increase following chemotherapy and impact overall survival (Sedláková et al. 2015). The human genome encodes 49 ABC genes, belonging to seven subfamilies, designated from A to G. The well-characterized ones are *ABCB1* (P-glycoprotein), *ABCC1 (MRP1)*, and *ABCG2 (BCRP)*, which are known to be upregulated in several cancers. Higher expression of *ABCB1* and *ABCC1* (at transcript and protein levels) was observed in EOC patients with shorter overall and progression-free survival. These patients also did not respond well to platinum-based therapies (Ohishi et al. 2002; Chen et al. 2014; Ween et al. 2015). Interestingly, knocking down or inhibiting the *ABCG2* activity sensitized the ovarian cancer cell lines to several drugs, including platinum-based drugs (Bram et al. 2009). Furthermore, recurrent cases undergoing chemotherapy were found to have higher levels of *ABCG2* (Zhu et al. 2015). One possible mechanism for such upregulation of these transporters might be the chemotherapeutic drugs themselves, as they were found to cause epigenetic modulation (demethylation) of the *ABCG2* promoter in ovarian carcinoma cell lines (Downie et al. 2005).

### 4.5.2 Increased Drug Inactivation

Drug metabolism is a crucial aspect of our body's ability to remain healthy. Most xenobiotics or drugs undergo chemical alteration and metabolic breakdown, usually through specialized enzymatic systems that are more easily excreted from the body. Drug-metabolizing enzymes (Phase I and/or II enzymes) primarily expressed in the liver and intestine metabolize most xenobiotic compounds. Resistant cancer cells hijack this machinery and over-express these enzymes to inactivate the intracellular drugs before they exert their cytotoxic effects. Some of these drug-metabolizing enzymes are briefly discussed below.

#### 4.5.2.1 CYP System

The cytochrome P450 enzymes are a group of mono-oxygenase enzymes that play a crucial role in the initial phase of drug metabolism by breaking down various therapeutic drugs and harmful environmental chemicals. Overexpression of *CYP1B1* is potentially involved in resistance to a range of anticancer drugs, particularly docetaxel, a chemotherapy agent for second-line therapy (Sawers et al. 2014). A study based on immunohistochemistry in primary ovarian cancer revealed that several P450 enzymes were present at significantly higher levels compared to the normal ovary. Additionally, CYP2S1 and P450 reductase expression were significantly increased in metastatic ovarian cancer compared to the primary tumor (Tong et al. 2019).

#### 4.5.2.2 Glutathione S-Transferase (GST)

GSTs are a cluster of enzymes involved in the process of detoxification, specifically in phase-II drug metabolism, where they facilitate the conjugation of glutathione (GSH) to toxic electrophilic xenobiotic compounds. Among the GSTs, GSTP1 is a well-known member that exhibits increased expression in ovarian cancer cell lines. Depletion of GSTP1 results in heightened sensitivity to cisplatin and carboplatin chemotherapy agents (Hamada et al. 1994). In the chemotherapy-sensitive group, the expression levels of GST-π were significantly lower compared to the chemotherapy-resistant group. Patients with elevated expression of GST-π displayed notably lower survival rates compared to those with low expression of the respective genes (Kakehi et al. 2015; Chornokur et al. 2015).

#### 4.5.2.3 Uridine Diphospho-glucuronosyltransferase (UGT) Superfamily

Uridine diphospho-glucuronosyltransferase (UGT) superfamily is also a part of phase-II enzymes that catalyze the conjugation of glucuronic acid to various substrates, including drugs and endogenous compounds (Milanesio et al. 2020). In the context of ovarian cancer, UGTs have been demonstrated to participate in the metabolic processes of specific chemotherapy drugs, such as irinotecan, thereby influencing the effectiveness of treatment (DiSilvestro et al. 2021).

### 4.5.3 Resistance to PARP Inhibitors and HR Pathways

As mentioned in previous sections, HGSOC is distinguished by genetic and epigenetic modifications occurring in the genes responsible for the Homologous Repair pathway, which encompasses *BRCA1/2*. As a consequence, the functionality of this pathway becomes compromised or defective (Rose et al. 2020). This leads to increased sensitivity to platinum agents and opens avenues for implementing therapies targeting poly (ADP-ribose) polymerase (PARPi) (Lin et al. 2019). PARPi works well due to the synthetic lethality produced and has been validated in clinical trials of patients with ovarian carcinoma carrying BRCA1/2 mutation. Olaparib, Rucaparib, and Niraparib have recently received approval for clinical utilization in the United States and Europe, specifically for treating platinum-sensitive ovarian cancers. These drugs have demonstrated the ability to enhance progression-free survival in affected individuals (Lin et al. 2019). Despite this progress, the acquirement of resistance to PARPi occurs in certain patients as the disease progresses. The development of drug resistance arises from the restoration of the Homologous Repair (HR) pathway, which subsequently repairs the DNA damage induced by PARP inhibitors and/or platinum-based chemotherapy. This restoration undermines the fundamental principle of synthetic lethality, promotes cell survival, and compromises the efficacy of the treatment (Kondrashova et al. 2017). The most common example of HRR restoration is frameshift or nonsense mutations (Barber et al. 2013) that restore the open reading frame of HR repair genes (BRCA reversion: *BRCA1/2, PALB2, RAD51C/D*) (Kondrashova et al. 2018; Castroviejo-Bermejo et al. 2018). The restoration of BRCA1/2 function can occur as a result of genetic events that reverse the frameshift caused by the initial mutation. This reversal reinstates the open reading frame (ORF), enabling the expression of a full-length protein. Additionally, genetic reversion of the inherited mutation can also lead to the restoration of a full-length, wild-type protein. These genetic events may occur due to selective pressure of exposure to cisplatin or PARPis. This resistance mechanism holds significant importance in patients with BRCA-mutated cancers undergoing platinum-based therapy. Among platinum-resistant BRCA-mutated high-grade serous ovarian cancers (HGSOCs), approximately 46% display tumor-specific secondary mutations that restore the open reading frame (ORF) of either *BRCA1* or *BRCA2*. Notably, a recent study examining the comprehensive genomic features of chemoresistant ovarian cancer has reported multiple instances of reversion events in *BRCA1/2* genes as a mechanism of platinum resistance (Nacson et al. 2018).

Demethylation of *BRCA1* (Dev et al. 2018) and *RAD51C* (Rottenberg et al. 2008) promoters also leads to the re-expression of the mRNAs and, eventually, the functional proteins. In primary HGSOC patients, there was a notable presence of promoter methylation and reduced expression of *BRCA1*. However, in patients with relapsed disease, the methylation of BRCA1 was diminished, and the expression of the *BRCA1* gene reached comparable levels to tumors with proficient homologous recombination mechanisms (Rose et al. 2020).

Recent investigations have provided evidence indicating that the conserved N- and C-terminal domains of BRCA play a pivotal role in the response to therapies targeting homologous recombination deficiency (HRD). Mutations occurring in the C-terminal domain, specifically the BRCT domain of BRCA1, result in the production of hypomorphic proteins that undergo protease-mediated degradation due to impaired folding. In the presence of PARP inhibitors, HSP90 stabilizes these truncated proteins, allowing them to effectively interact with other protein complexes, forming RAD51 foci. Consequently, this process contributes to the development of resistance to PARP inhibitors and cisplatin (Amaravadi et al. 2011). Additionally, removing the end resection barrier due to the loss of *53BP1* (Wang and Wu 2014) or proteins comprising the Shieldin complex (Bishnu et al. 2021) leads to restoring HR independent of *BRCA1*. Finally, most of the PARPi are substrates of several drug efflux transporters, and an increase in efflux, notably through ABCB1 upregulation, reduces intracellular PARPi concentrations, inducing resistance (Lu et al. 2008).

### 4.5.4 Deregulated Autophagy

Autophagy is a cellular mechanism characterized by the breakdown and recycling of cellular components, such as damaged or unneeded proteins and organelles. Deregulation of autophagy can occur in cancer cells and contribute to their survival and resistance to therapy. Several genes known to be altered in ovarian cancer like (*PTEN, TP53, LC3, DRAM*) and ERK and AKT-mTOR pathways dysregulate autophagy. Interestingly, autophagy plays a complex and dual role in cancer (Zhang et al. 2015a). On one hand, autophagy can act as a tumour suppressor mechanism by removing damaged or abnormal cells, including cancer cells. This can help to prevent the formation and progression of tumours. On the other hand, autophagy can act as a pro-survival mechanism in cancer cells, helping them to resist stress, including chemotherapy and radiation therapy.

Cancer cells can use autophagy to degrade and recycle cellular components, such as proteins and organelles, to maintain energy levels and provide a survival advantage. In ovarian cancer, autophagy is upregulated in vitro and in-vivo studies, which can lead to resistance to chemotherapy and poor prognosis. This upregulated autophagy might result from cisplatin-mediated extracellular signal-regulated kinase (ERK) activation. Indeed, the autophagy flux decreased and increased the sensitivity of cisplatin in several ovarian cancer cell lines after inhibiting the activity of ERK (Xiao et al. 2016). Mechanistically, inhibition of ERK activity led to a decrease in the two proteins necessary for autophagosome–lysosome fusion, i.e., UVRAG and RAB7, which led to stalling of autophagic flux necessary to tolerate the chemotherapeutic insults (Kuo et al. 2019). It has also been demonstrated that the ARH1 protein has the ability to enhance autophagy by inhibiting the mTOR pathway, as well as trigger a state of dormancy in ovarian cancer cells that relies on autophagy (Dai et al. 2014). The initiation of autophagy, facilitated by either thioredoxin domain containing 17 (TXNDC17) (Niu et al. 2019) or *YAP* activation (An et al. 2018) has been observed to enhance cisplatin resistance in diverse ovarian cancer cell lines.

Metastasis, which involves the spread of cancer cells from the original site, is a crucial process in tumour progression. In ovarian cancer (OC), peritoneal seeding is the most common route of metastasis, and research has shown that autophagy contributes to this process. Autophagy plays a significant role in the selection of metastatic cells capable of thriving in the peritoneum (Zhang et al. 2015b). In the study conducted by Kuo et al. in 2019, initially, EOC cells were injected into mice, and peritoneal metastases were collected, cultured, and subsequently implanted into new hosts. This selection process was repeated three times to generate sublines. The derived sublines originating from the metastatic sites exhibited not only an enrichment of autophagy-related genes but also demonstrated an increased ability to induce autophagy in response to stressors such as serum starvation or treatment with cisplatin compared to the parental cell lines. The growth of these sublines was significantly impeded upon inhibition of autophagy through ATG5 knockdown. These findings provide evidence that the upregulation of autophagy confers a survival advantage to metastatic cells, suggesting that targeting autophagy could hold clinical benefits for addressing peritoneal metastases in ovarian cancer.

Furthermore, the examination of autophagy markers' expression presents promise as a prognostic tool in epithelial ovarian cancer (EOC). In a study, the expression of ATG9A, a crucial protein involved in autophagosome formation, was assessed in tissue samples collected from 160 patients with ovarian adenocarcinoma. The results revealed a significant correlation between high ATG9A expression and advanced clinical stage as well as disease relapse, indicating its potential as a prognostic indicator in EOC (Ricci et al. 2019). Another recent study developed a novel scoring system utilizing a panel of 43 ATG genes to accurately predict outcomes for advanced HGSOC patients after chemotherapy. By analyzing overall survival and recurrence-free survival (RFS) using a training dataset from TCGA data, the authors found that a higher ATG score was associated with shorter RFS following first-line treatment.

Moreover, an elevated ATG score was found to be associated with unfavorable overall survival, independent of residual disease status, which is a critical clinical factor for predicting patient outcomes following treatment (Giacomini et al. 2020). Investigating the connection between the expression of autophagy-related genes and the clinical outcome of individuals with ovarian serous cystadenocarcinoma, the study identified a risk signature related to autophagy, consisting of eight genes: *BLOC1S1, IL24, NRG4, PDK4, PEX3, PRKG1, SIRT2*, and *WDR45L* (Catanzaro et al. 2015).

The strategic targeting of autophagy holds promise as an appealing strategy for the clinical management of ovarian cancer, offering potential solutions for overcoming chemoresistance, managing metastasis, and serving as a prognostic biomarker. Nevertheless, considering the intricate involvement of autophagy in regulating diverse aspects of tumor biology, a deeper comprehension of the molecular pathways that govern autophagy in epithelial ovarian cancer (EOC) is imperative.

### 4.5.5 Altered Metabolism

Altered metabolism is a hallmark of cancer cells, which predominantly rely on glucose metabolism, fatty acid synthesis, and glutaminolysis. Such altered metabolic rewiring enables cancer cells to attain enhanced proliferation and survival benefits.

#### 4.5.5.1 Glucose Metabolism

Cancer cells have a higher rate of glucose uptake and use than normal cells, a process known as the Warburg effect. This increased glucose metabolism allows cancer cells to generate the energy they need to grow and proliferate. Although less efficient in generating ATP than oxidative phosphorylation (OXPHOS), the precursors produced in glycolysis are usually channeled downstream into several anabolic pathways required by continuously proliferating cancer cells. Treatment with 2-deoxyglucose (2DG), a non-metabolizable glucose analog, sensitized EOC cells to cisplatin (Zhang et al. 2015c). However, OXPHOS capabilities are also reported in cancer cells, highlighting their ability to adapt to the ever-changing microenvironment of the tumour. Enhanced OXPHOS is observed in platinum resistant PDX models of (Li et al. 2017). The Pentose Phosphate Pathway (PPP) also metabolizes glucose to produce NADPH and ribose 5-phosphate (R5P), which are involved in various biosynthetic reduction reactions and also regenerates reduced glutathione (GSH) that protects cells against oxidative stress. Since platinum drugs sequester GSH and generate ROS, upregulation of PPP is associated with increased platinum resistance (Iida et al. 2012). Interestingly, in a cellular model of EOC resistance, an increase in G6PD enzymatic and PPP activity was observed, and inhibition of these activities led to increased platinum sensitivity (Pelletier et al. 2012).

#### 4.5.5.2 Fatty Acid Metabolism

Lipid reserves are rich energy sources, therefore, increased lipid synthesis is a means to meet the ever-increasing energy expenditure inside a cancer cell. Lipids are also the source of membrane biosynthesis. Unsaturated lipids are high in ovarian cancer, which induces stemness, whereas lipid desaturation impairs cancer stemness and tumour initiation (Yang et al. 2014; Hudson et al. 2016). Clear Cell Ovarian Carcinoma, known to be upfront resistant to various chemotherapeutic drugs, shows a characteristic increase in cellular glycogen storage (Obrist et al. 2018). In addition, recent studies have shown that hypoxic conditions are known to induce glycogen synthesis through HIF1-$\alpha$, and such conditions exist in the core of the solid tumour microenvironment (Baekelandt et al. 1999). All these observations indicate that a particular metabolic phenotype might enhance chemoresistance in EOC.

#### 4.5.5.3 Altered Glutamine Metabolism

Despite being one of the most abundant amino acids in the blood, glutamine shortage is observed in the core of the tumour, thereby rendering glutamine an essential amino acid for the tumour cells. Tumour cells mostly use glutamine to produce energy via OXPHOS, which is essential for the highly proliferating metastatic ovarian cancer cells. Upregulated glutamine metabolism was found in highly invasive ovarian cancer cells, which were addicted to glutamine compared to noninvasive ovarian cancer cell lines (Binju et al. 2019). Such dependency might result from the overexpression of glutamine transporter ASCT2 and glutaminase through the activation of STAT3 or c-MYC (Zervantonakis et al. 2017). Platinum resistant endometrioid EOC cell line was found to depend on glutamine metabolism, where the glutamine was principally used for nucleotide biosynthesis (Lavarino et al. 2000).

### 4.5.6 Evasion of Apoptosis

Apoptosis, a biological phenomenon occurring in multicellular organisms, plays a vital role in development and homeostasis by ensuring the maintenance of cellular populations. Platinum-derived drugs like cisplatin or carboplatin are alkylating agents that produce DNA adducts/crosslinks, and such DNA damage leads to the activation of mitochondrial-mediated apoptosis (Reles et al. 2001). The cellular fate of survival or death is determined by the balance of anti-apoptotic proteins such as BCL2 and BCL-XL, and pro-apoptotic proteins such as PUMA and BAX (Bitler et al. 2017). Ablation of pro-survival-related proteins in ovarian cancer cell lines increases cisplatin sensitivity (Fraser et al. 2003). The upregulation of the PI3K/mTOR axis suppresses the activity of pro-apoptotic BCL-2 family proteins, thereby promoting the development of resistance (Fraser et al. 2003). Inactivation and mutations in p53, a master regulator of apoptotic pathway, are known to increase resistance to various chemotherapeutic agents (Hajra et al. 2008). Thus, overexpression of wild-type p53 protein in the A2780/CP resistant cell line enhanced sensitivity to platinum-based drugs (Mor et al. 2008). Another driver for evading apoptosis are *ARID1A* mutations, which upregulate *HDAC6* that epigenetically inactivates the apoptosis inducing functions of p53 (Stewart et al. 2011). Besides the above-mentioned studies, a large body of literature described various mechanisms associated with apoptotic pathway members and chemoresistance in EOC, which are well presented in many reviews (Chen et al. 2013; van der Gun et al. 2010; Choi et al. 2015).

### 4.5.7 Cancer Stem Cells

Cancer Stem cells are a small population of cells within the solid tumour that share several properties of stem cells and are capable of self-renewal and repopulating the tumours following therapy. These cells can be isolated based on several surface markers like CD133+ (Yamanoi et al. 2019), CD44+/CD117+ (Tao et al. 2018), and Epithelial Cell Adhesion Molecule (EpCAM) (Radpour 2017), functional markers like aldehyde dehydrogenase (ALDH) + positivity (Bellio et al. 2019) and Side population (Begicevic and Falasca 2017; Zinzi et al. 2014). They can self-renew and differentiate to produce varied cell phenotypes, leading to tumour heterogeneity. Several studies have positively correlated CSCs with the prognosis of EOC (Nagaraj et al. 2015). Various mechanisms such as faster DNA repair, altered efflux/influx transporters, activated signaling pathways etc. have been reported to explain their intrinsic chemoresistant nature. Enhanced homologous recombination repair efficiency was found in CSCs of PARPi-resistant ovarian cancer patients (Nowicki et al. 2021; Son and Moon 2010). Another mechanism that avoids drug penetration is the proficient drug-effluxing abilities of upregulated drug-efflux transporters belonging to the ABC gene family (Gao et al. 2014; Dou and Zhang 2021). In addition, drug-metabolizing enzymes like ALDH are over-expressed in CSCs. These enzymes aid in the oxidation of aldehydes to carboxylic acids, thereby protecting DNA. These enzymes also activate several stemness-maintaining pathways like Wnt/b-Catenin or hedgehog pathways (Boyerinas et al. 2012; Sun et al. 2013). Another well-known mechanism associated with CSCs is epithelial to mesenchymal transition (EMT), wherein a cell loses its epithelial nature and takes up mesenchymal phenotype becoming highly invasive, motile, and malignant (Yang et al. 2012a). The transition seems to be controlled by the transforming growth factor (TGF-B) pathway as there was a strong positive correlation was observed between TFGB-mediated EMT and chemoresistance (Zhou et al. 2014).

### 4.5.8 Role of miRNAs and lncRNA

MicroRNAs (miRNAs) are small, non-coding, highly conserved RNAs ranging from about 20–25 nucleotides in size in the post-transcriptional negative regulation of gene expression. They pair with complementary sequences within mRNA molecules and trigger the following processes to silence these mRNA molecules: (1) Cleavage of the mRNA strand into two pieces by RNA-induced silencing complex (RISC) complex, (2) shortening the poly (A) tail thereby destabilizing the mRNA (3) decreasing the efficiency of translation by ribosomes. Compelling evidence suggests that deregulation of miRNA expression is implicated in cancer through various mechanisms, including mutations, changes in the copy numbers, chromosomal rearrangements, epigenetic modification, defects in the miRNA biogenesis pathway, or regulation via transcriptional factors.

Depending on their function, miRNAs may be classified as oncogenes or tumour suppressors, regulating different hallmarks of cancer. The role of microRNAs (miRNAs) in conferring chemoresistance in ovarian cancer cells has been extensively documented in numerous studies. In ovarian cancer, the dysregulation of two miRNA families, miR-200 and let-7, is frequently observed and has been linked to epithelial-to-mesenchymal transition (EMT), invasion, and chemoresistance. An example is the targeting of PTEN (a tumor suppressor) by miR-4461, which promotes enhanced cell growth and metastasis in ovarian cancer cells, indicating the oncogenic role of miR-4461 (Li et al. 2014). In many studies, miRNAs have been shown to be involved in drug resistance. They may target genes involved in drug resistance and/or affect genes that prevent cancer spread. Studies have demonstrated that specific miRNAs, including let-7 (Yang et al. 2012b), miRNA-9 (Statello et al. 2021), miRNA-370 (Gao et al. 2020), and miRNA-449a (Wang et al. 2019) play a role in conferring cisplatin sensitivity to EOC cells. While others like miRNA-21 (Xu et al. 2018a) and miRNA-93 (Liu et al. 2018), promote resistance to cisplatin in EOC cells. Intriguingly, miRNAs target a group of genes rather than a single gene, complicating the understanding of their mechanism of action.

Conversely, long non-coding RNAs (lncRNAs) are non-coding RNA transcripts that are highly conserved and exceed 200 nucleotides in length. These transcripts regulate gene expression by modulating various processes such as chromatin remodeling, transcription, and alternative splicing and can generate miRNAs or produce short biologically active peptides (Liu et al. 2015). According to their mechanism of action, lncRNAs are classified into four categories (Wang et al. 2015). LncRNAs can exhibit various functional roles, serving as signals that indicate functional biological conditions in a spatiotemporal manner, responding to developmental cues, cellular signals, and other stimuli. They can also function as decoys, interfering or competing with other RNAs or proteins that bind to DNA, thereby repressing transcription. Furthermore, lncRNAs can act as guides by binding to proteins and facilitating their transport to chromatin and other specific targets, thereby inducing changes in gene expression of neighboring or distantly located genes. Additionally, lncRNAs can function as scaffolds, serving as central platforms for assembling different effector molecules. Through these assembled complexes, specific signaling components and outputs can be selectively utilized to initiate corresponding biological functions, such as transcriptional repression or activation.

LncRNAs have been clinically recognized for their involvement in ovarian cancer, as they play a significant role in the development of treatment resistance, which ultimately contributes to disease relapse and unfavorable patient prognosis (Haibe et al. 2020). LncRNA EBIC was found to play a role in promoting cisplatin resistance in epithelial ovarian cancer (EOC) cells by activating the WNT/beta-catenin signaling pathway (Tao and Wu 2021). Similarly, HOX antisense intergenic RNA (HOTAIR) was positively associated with chemoresistance and contributed to increased chemoresistance by activating the same pathway in EOC cells. Inhibition of lncRNA HOTAIR had a negative impact on drug-resistant EOC cell activity and led to suppression of *HOXA7* expression (Christie et al. 2019). Moreover, overexpression of lncRNA PVT1 enhanced cisplatin resistance in EOC cells by regulating key genes involved in apoptotic pathways, including TGF-beta1, p-Smad4, and caspase-3 (Rottenberg et al. 2008). Additionally, lncRNA UCA1 induced cisplatin resistance in EOC cells with the involvement of SRPK1 (Gogola et al. 2018).

## 4.6 An Insight Into the Upcoming Therapies and Clinical Trials

### 4.6.1 Overcoming Bevacizumab Resistance Through Peptide Fusion Protein Trebananib

Resistance to anti-angiogenic therapy is a significant challenge in cancer treatment as a substantial number of patients are intrinsically resistant to anti-angiogenic drugs and show only minimal or no response (Haibe et al. 2020). Mechanisms of acquired resistance in EOC (Tao and Wu 2021) include upregulation of drug efflux proteins including *ABCB1* (Christie et al. 2019; Rottenberg et al. 2008), changes in PAR metabolism (Gogola et al. 2018; Pettitt et al. 2018) and BRCA reversion leading to restoration of the HR pathway (Christie et al. 2017; Sakai et al. 2008; Patch et al. 2015). Redundancy of angiogenic signals in tumours pose another challenge as bevacizumab treatment leads to the activation of alternate pathways involving factors other than VEGF (Bergers and Hanahan 2008).

Trebananib (AMG386) is a peptide fusion protein or a pepti-body that targets the ANG1/ANG2/TIE2 pathway. It blocks the interaction of angiopoetins, ANG1 and ANG2 to the Tie2 receptor and inhibits angiogenesis via a VEGF independent pathway (Liontos et al. 2014). In a double blind, phase III trial (TRINOVA-1), combination of Trebananib was investigated in combination with paclitaxel in ovarian cancer patients with platinum resistant recurrent disease. Patients were given weekly paclitaxel along with weekly Trebananib or placebo. Progression free survival in the Trebananib plus paclitaxel arm significantly improved (7.2 vs 5.4 months) compared to the paclitaxel only arm (Monk et al. 2014). However, no improvement in the overall survival was observed. In another double blinded phase III trial (AGO-OVAR 18), the efficacy of combination therapy of carboplatin and paclitaxel along with subsequent weekly Trebananib is being investigated (Vergote et al. 2019).

## 4.6.2 Emerging Targeted Therapies

Currently in ovarian cancer, potential therapeutic targets include hormone receptors, angiogenesis, Homologous recombination deficiency, certain tumour specific signaling pathways and immunological factors. For Gynecological malignancies, bevacizumab, pembrolizumab, olaparib, niraparib, and rucaparib have been FDA approved for selected patients with recurrent metastatic disease. However, these inhibitors are applicable to only a small subset of patients with defined genetic alteration and a significant proportion of patients still need more precise therapeutic options. Some of the upcoming targeted therapies are as follows:

### 4.6.2.1 Anti-angiogenic agents

**Pazopanib** Efficacy of single agent Pazopanib, an oral tyrosine kinase inhibitor (TKI) of VEGFR 1, 2 & 3, PDGF receptor (PDGFR) α & β, and c-KIT, was evaluated in a phase III, randomized, two arm placebo controlled, double blinded trial (AGO-OVAR16), in 940 patients with stage II-IV ovarian or primary peritoneal cancer, post first line of chemotherapy (Vergote et al. 2018). Pazopanib maintenance therapy improved progression free survival by 5.6 months compared to the placebo control but no difference was observed in the overall survival (Floquet et al. 2015).

**Nintedanib** Nintedanib, another oral TKI of VEGFR 1,2,3, FGFR1,2,3 and PDGFR α, β, showed an improvement in PFS in a phase II trial conducted in patients with platinum sensitive, recurrent EOC (16.3% vs 5%, $p = 0.06$) (Ledermann et al. 2011). Subsequently, efficacy of Nintedanib in combination with first line chemotherapy is being investigated in another phase III trial (AGO OVAR12-NCT01015118) for newly diagnosed patients with EOC (stages IIb-IV) (Ray-Coquard et al. 2020).

**Cediranib** Cediranib is a TKI of VEGFR-1/-2/-3 and, c-KIT. In a phase III, three-arm, double blinded, placebo-controlled trial, ICON6, combination of cediranib with chemotherapy as well as maintenance treatment is being investigated in patients with platinum-sensitive recurrent EOC. Though the median PFS in the cediranib maintenance arm was found 2.3 months longer compared to placebo, the trial is still continuing and patients are being followed up for the assessment of overall survival, which is the secondary endpoint of the study (Ledermann et al. 2016).

**Apatinib** Apatinib, an oral small molecule TKI, binds to the ATP-binding site of the VEGFR-2 and has been investigated as a salvage treatment for advanced stage solid tumours such as gastric cancer. Single agent apatinib, in a phase II trial, in patients with advanced recurrent EOC maintained stable disease in 68.9% of patients with a median OS of 14.5 months and PFS of 5.1 months, compared to placebo (Miao et al. 2018).

#### 4.6.2.2 Targeting the DNA Repair Pathway

Even though debulking surgery followed by platinum-taxol therapy serves as the first line of treatment in ovarian cancer, majority of the patients show disease relapse and chemoresistance after subsequent treatments. The HRD phenotype makes the cancer cells not only susceptible to platinum treatment but also enhances the therapeutic efficacy of poly- (ADP-ribose) polymerase (PARP) inhibitors. This therapeutic strategy has been proven to be successful in HGSOC patients (Fong et al. 2010). As discussed earlier, PARP enzymes belong to a group of proteins which are involved in base excision repair. PARP1 and PARP2 detect and catalyze the repair of single-strand DNA breaks (Morales et al. 2014). Olaparib, niraparib and rucaparib are FDA-approved PARPis which have showed immense promise as maintenance therapy in randomized Phase III clinical trial of ovarian cancer (NCT01844986, NCT02655016, NCT01968213). A meta-analysis of four qualified and relevant randomized clinical trials showed that PARPis are significantly beneficial as maintenance therapy for both platinum-sensitive relapsed ovarian cancer and advanced ovarian cancer, irrespective of *BRCA* mutation or HRD status (Lin et al. 2019). However, a detail retrospective analysis conducted by Gulia et al., on these four clinical trials suggest that PARPis are effective in HRD-positive tumours as is observed in a significant change in PFS [45.7 months (PARPi treated) vs 17 months (placebo)]. They did not observe a remarkable difference in the HRD-negative tumours (Gulia et al. 2022). However, their efficacy in non-HRD patients has been attributed to the ability of PARPis to trigger ferroptosis through the repression of *SLC7A11*, a cystine transporter, in a p53-dependent manner (Hong et al. 2021). Since most cases of ovarian cancer gradually develop resistance to these inhibitors, current research is majorly focused on extending the utility of PARPis. A phase II clinical study involving the combinatorial administration of cediranib (anti-angiogenic) plus olaparib demonstrated an improved median PFS of 16.5 months in comparison to 8.3 months for only olaparib treatment (Liu et al. 2019). Recently, FDA has also approved the combination of olaparib and bevacizumab as maintenance therapy for newly diagnosed patients with stage III and IV ovarian cancer based on the results of the PAOLA-1 trial (Arora et al. 2021). The rationale behind using an anti-angiogenic agent in combination with PARPi is that angiogenesis inhibition leads to induction of hypoxia which is crucial in downregulating BRCA1/2 and RAD51 thereby disrupting HRR (Lim et al. 2014; Bindra et al. 2007) . Angiogenesis was observed to be severely disrupted in PARP1 knockout mice which can be due to VEGF-A downregulation due to PARP1 absence (Tentori et al. 2007; Wei et al. 2016). Additionally, multiple clinical studies to evaluate the efficacy of such combination are underway which pair PARPis with ATR inhibitors (NCT03462342), WEE-1 inhibitors (NCT03579316), AKT inhibitors (NCT02338622, NCT02208375;), MEK inhibitors (NCT03162627), and mTORC1/2 inhibitors (NCT02208375). A study has also reported the synergistic effect of choloroquine and panobinostat, a histone deacetylase inhibitor, by targeting DNA repair pathway. Choloroquine was found to induce double strand breaks due an increase in the percentage of cells showing $\gamma H_2AX$ foci whereas panobinostat was found to inhibit the correct recruitment of RAD51 to the double stranded breaks, bringing about this synergism (Ovejero-Sánchez et al. 2021).

#### 4.6.2.3 Receptor Tyrosine Kinase Inhibitors

**Sunitinib**

Sunitinib is a small molecule multi-target receptor Tyrosine kinase inhibitor (Mendel et al. 2003) approved for the treatment of gastrointestinal stromal tumours and renal clear cell carcinoma (Motzer et al. 2006). It targets PDGFs as well as VEGFRs and cKIT. It is currently being investigated in a phase 2 clinical trial (NCT01824615) for platinum resistant/recurrent ovarian clear cell carcinoma patients who have received prior platinum-based therapy.

#### 4.6.2.4 Epidermal Growth Factor Receptor- Family Targeted Inhibitors

Amplifications in the EGFR family of receptors (EGFR, ERBB2, ERBB3, AND ERBB4) have been implicated in EOC (Wilken et al. 2012). However, targeted anti-EGFR therapy has only shown limited clinical efficacy in EOC patients. Few of the anti EGFR agents currently in clinical trials are as follows:

**Gefitinib** The maximum tolerated dose and response rate of Gefitinib in combination with topotecan is currently being investigated in patients with relapsed, platinum-resistant, ovarian, peritoneal or fallopian tube, EGFR Positive cancers, in a Phase I/II clinical trial (NCT00317772).

**Pertuzumab** Efficacy of Pertuzumab in combination with chemotherapy was investigated in a Two-Part, Randomized Phase III, Double-Blinded placebo controlled clinical trial(AGO-OVAR 2.29/ ENGOT-ov14/PENELOPE) in patients with recurrent Platinum-Resistant Epithelial ovarian cancer and low HER3 mRNA Expression but no improvement in the overall survival was observed (Kurzeder et al. 2016), however, pertuzumab in combination with gemcitabine or paclitaxel did improve the PFS compared to placebo plus chemotherapy (4.3 months vs 2.6 months), indicating the need for further validation of pertuzumab in therapy resistant ovarian cancer.

#### 4.6.2.5 MEK Inhibitors

The MEK inhibitor, selumetinib, has shown benefit in patients with recurrent LGSOC (NCT00551070) and dual targeting of PI3K/ AKT and MAPK pathway may an attractive strategy in the future (Farley et al. 2013).

#### 4.6.2.6 PI3K Inhibitors

**Alpelisib** Alpelisib is a potent PIK3CA specific inhibitor, recently approved by the FDA for the treating HER2 negative, hormone receptor positive, PIK3CA mutated, advanced or metastatic breast cancer, in combination with fulvestrant (André et al. 2019). Alpelisib is a potential therapeutic target in epithelial ovarian cancer which report alterations in the PIK3CA gene, such as clear cell and Endometroid carcinomas (activating mutations) and High Grade Serous Ovarian Carcinoma (copy number amplifications). In a phase 1b trial, clinical evidence of synergism of Alpelisib and Olaparib in combination was reported in Epithelial ovarian cancer patients irrespective of their *BRCA* status (Konstantinopoulos et al. 2019).

### 4.6.3 Strategies to Develop Additional Synthetic Lethal Therapies

#### 4.6.3.1 Wee1 Inhibition

Wee1 is a serine/threonine kinase that regulates the G2/M checkpoint transition. It induces an arrest in the G2/M phase by phosphorylating CDK1 at Tyrosine-15 which leads to CDK1 inhibition and prevents the entry of cell into mitosis (Watanabe et al. 1995). Inhibition of WEE1 also leads to high CDK1 activity resulting in G2M progression without adequate DNA repair. Thus, the cell relies solely on the G1/S checkpoint arrest which is regulated by wild-type p53 (Wang et al. 2004). In case of ovarian cancer, where there is a high prevalence of *TP53* mutations, WEE1 inhibition disrupts both G1/S and G2M checkpoints, resulting in synthetic lethality when treated with cytotoxic agents. Adavosertib (AZD1775), a potent, small molecule inhibitor of Wee1 kinase, synergistically enhances the activity of various DNA damaging chemotherapeutics such as carboplatin, cisplatin and gemcitabine, in p53-deficient cells (Moore et al. 2022). In a phase 1 study, AZD1775 in combination with carboplatin was given to *TP53* mutated, platinum resistant or refractory ovarian cancer patients, where the ORR was 43% with a median PFS of 12.6 months in the combination group (Leijen et al. 2016).

#### 4.6.3.2 CHK1 Inhibitors

CHK1 is a cell cycle checkpoint kinase activated by ATM/ATR in response to DNA damage. CHK1 phosphorylates and degrades its targets CDC25A/B/C resulting in inhibition of CDKs and subsequent S and G2M phase arrest, CHK1 is known to also activate WEE1 kinase, ultimately leading to CDK1 inhibition. CHK1 inhibitors are highly synergistic with replication-dependent DNA damaging agents. In a phase 2 trial, CHK1 inhibitor, Prexasertib efficacy is being investigated in platinum resistant or refractory high-grade serous ovarian cancer patients (Konstantinopoulos et al. 2022). In another phase 1 study, prexasertib in combination with Olaparib is being evaluated in EOC patients with de novo or acquired PARP inhibitor resistance. Findings suggest that CHK1 inhibitor sensitized the tumours to PARP inhibition (Do et al. 2021).

### 4.6.4 Sequential and Combinatorial Approaches Using Chemotherapeutics and Molecular Targeted Agents

Combinatorial and sequential drug therapies both have their place in the treatment of cancer. Often it is found that chemotherapeutic agents administered in combination are more toxic than either drug monotherapy, ultimately, compromising the quality of life of the patient and thus, having little or no clinical benefit. However, therapy combinations, where chemotherapeutic agents are used in combination with inhibitors/targeted therapies are relatively less toxic and much well tolerated by patients with manageable safety profiles. Such therapies are preferred in patients when the tumour burden is high.

Sequential therapy on the other hand, allows for the optimal delivery of both drugs, without high toxicity in the patients and without compromising quality of life. It is the preferred approach in frail as well as elderly patients, who are unable to tolerate combinatorial treatment. It is also preferred in patient with slow growing tumours.

The standard therapeutic regimen in ovarian cancer includes a combination of platinum with paclitaxel, with or without Bevacizumab for maintenance therapy. Some of the upcoming drug combination currently being explored in various drug trials are as follows

**Adavosertib plus Gemcitabine** Adavosertib (AZD1775) is a small molecule inhibitor of the Wee1 kinase that has previously been reported to show synergy with gemcitabine in Phase I clinical trials. In a double blinded, randomized phase II clinical trial of women with recurrent platinum-refractory or platinum-resistant high-grade serous ovarian cancer, patients were either given oral adavosertib or identical placebo along with weekly intravenous gemcitabine. Their findings showed that progression free survival significantly improved in the adavosertib plus gemcitabine group (median 4.6 months) compared to the placebo plus gemcitabine group (3 months) (Lheureux et al. 2021).

**Adavosertib plus Carboplatin and Paclitaxel** A phase II randomized, double blinded, placebo controlled, clinical trial with 121 p53-mutated platinum sensitive EOC patients receiving adavosertib in combination with carboplatin, reported that addition of adavosertib to standard chemotherapy improved progression free survival (NCT01357161) (Oza et al. 2020). Another phase II study with platinum resistant ovarian cancer reported that addition of adavosertib to standard chemotherapy (showed an ORR of 67% (carboplatin plus adavosertib arm) with a PFS of 12 months (Moore et al. 2022).

**Pembrolizumab (Anti-PD-1) and AMG386 (Trebananib)** Pembrolizumab, a selective humanized monoclonal antibody against the PD1 receptor, prevents the binding and activation of the PDL1 and PDL2 ligands. In a phase Ib trial of patients with advanced solid tumours, including ovarian cancer patients that have relapsed within 6 months of platinum therapy (platinum resistant ovarian cancer), the safety and synergy of combination of Pembrolizumab with AMG386 is being investigated. The dose limiting toxicities as well as the overall response rate, progression free survival and the overall survival of these patients will be assessed (Rahma et al. 2019).

**Paclitaxel plus Trebananib** In a randomized, multicenter, double-blind, placebo-controlled phase 3 trial, women with recurrent partially platinum sensitive or resistant epithelial ovarian cancer were treated with intravenous paclitaxel along with weekly Trebanabib (IV). Findings show that median PFS was significantly longer in the trebananib group compared to placebo (7.2 months vs 5.4 months) and typical adverse effects associated with anti-angiogenic treatment were not prominent (Monk et al. 2016). Thus, non-VEGF, antiangiogenic treatment showed efficacy in ovarian cancer patients and should be explored further with other treatment combinations as well.

**Pazopanib plus Oral Cyclophosphamide** Efficacy of oral cyclophosphamide in combination with Pazopanib was evaluated in platinum resistant relapsed ovarian cancer patients treated with chemotherapy in a phase I/II clinical trial (PACOVAR). A median PFS of 8.3 months and OS of 24.95 months was observed (Dinkic et al. 2017). Furthermore, a retrospective analysis was conducted with 20 patients (17–platinum resistant and 3-platinum refractory) to study the efficacy of this treatment regimen, where the median PFS and OS was observed to be 5.5 and 9.5 months, respectively (Gulia et al. 2020).

**Apatinib plus Etoposide** In a phase 2 single arm prospective study (AEROC), Apatinib in combination with oral etoposide in patients with platinum-refractory and platinum-resistant ovarian cancer showed objective responses in 54% of the patients (Lan et al. 2018).

**Sintilimab plus Bevacizumab** Sintilimab, a humanized IgG4 monoclonal antibody against the programmed cell death receptor-1 (PD-1) blocks the interaction of PD-1 with the PD-L1 and PD-L2 ligands, and restores the endogenous antitumour T-cell response (Zhang et al. 2018). Bevacizumab and PD1 inhibitors have both shown efficacy in preclinical trials for OCCC both as single agents and in combination (Ogasawara et al. 2020; Disis et al. 2019; Hamanishi et al. 2015; Amit et al. 2017). Currently, an ongoing INOVA trial is investigating the efficacy of Sintilimab and Bevacizumab in combination, in patients with Recurrent/persistent Ovarian clear cell carcinoma who have previously received at least one platinum-based chemotherapy (Li et al. 2022).

### 4.6.5 Targeting Immune Evasion Mechanisms

#### 4.6.5.1 Adoptive T-Cell Therapies

**CAR-T Cell Therapy (Targeting MUC16, Folate Receptor α)**

Immune evasion or the ability of cancer cells to escape elimination triggered by active immune cells has shed light into the ambiguous role played by the host immune system in enabling or eradicating tumour development (Hanahan and Weinberg 2011). Incipient neoplasia has tactfully developed mechanisms to avoid the cytotoxic effects of the host immune surveillances. Therefore, the concept of immunotherapy has emerged from the prevalent interactions between the tumour cells and the immune system. Chimeric Antigen Receptors (CARs) T-cells which were first designed in 1989 have become a much-needed tumour specific alternative to the conventional chemotherapy and radiation therapy (Gross et al. 1989). Autologous T-cells collected from the patients are genetically modified ex-vivo to target tumour specific antigens after which they are expanded in culture and infused back (Miliotou and Papadopoulou 2018). Inside the body, they have the ability to recognize carbohydrates, glycoprotein and glycolipid structures which are typically expressed by tumour cells and not presented by major histocompatibility complex (MHC) (Schmidt-Wolf et al. 1991). Yescarta™ and Kymriah™ became the first two CAR-T therapies against B-cell acute lymphoblastic leukemia and B-cell lymphoma respectively to gain FDA approval in 2017 (Seimetz et al. 2019). Majority of these therapies especially for solid tumours are currently under multiple clinical and pre-clinical trials. Kershaw et al., conducted the first clinical study involving the use of first-generation CAR-T cells in ovarian cancer therapy. However, no evidence of antitumorigenic activity was observed in any patient. This poor trafficking of T-cells was attributed to their retention in lungs, liver, and spleen. It was also argued that these cells showed low persistence in patients as their numbers were initially high but plummeted quickly by day 5 post infusion and were undetectable beyond 3 weeks (Kershaw et al. 2006). In 2011, Song et al. developed a second-generation CAR-T cell therapy wherein the T cells expressed folate receptor-α (FRα) targeting CAR along with a CD137 signaling domain. These modified CAR-T cells showed significant tumour regression and no tumour related mortality along with enhanced tumour homing in human ovarian cancer xenograft models (Song et al. 2011). Apart from targeting FRα, tandem PD1-antiMUC16 dual-CAR have also been developed which target mucin 16 (MUC16, CA125) and programmed cell death-1 (PD-1). The mean survival time of dual CAR-T group reached to $80.6 \pm 10.33$ days vs $19.8 \pm 2.14$ days in the control group showing promising therapeutic efficacy in-vivo (Li and Wang 2020). Furthermore, a dose escalation phase I clinical trial is currently underway to determine the safety and preliminary efficacy of autologous IL-12 secreting, MUC-16$^{ecto}$ CAR-T cells in patients with recurrent MUC16$^{ecto+}$ ovarian, fallopian tube, or primary peritoneal cancer (Koneru et al. 2015). The presence of IL-12 leads to the decrease in the immune suppressive macrophages in the tumour microenvironment. Despite an enhanced therapeutic response over conventional treatment regimen in hematological malignancies, similar efficacy is not observed in case of solid tumours along with an added risk of cytokine release storm (CRS) and immunologic toxicities (Wu et al. 2021).

### TCR- Therapy (Targeting MAGE-A4, NY-ESO-1)

T-cell receptor-based (TCR) therapy also involves genetic modification of autologous T lymphocytes ex-vivo such that they express TCRs with enhanced binding affinities, which are then injected back into the patients to fight cancer by targeting tumour cells. However, their mechanism of recognizing antigen is markedly different from the CAR-T cells. TCR is a heterodimer that is made up of two different transmembrane polypeptide chains: α and β. The α and β peptide chains of TCRs are involved in the identification and binding of T lymphocytes to antigens presented by MHC molecules (Xu et al. 2018b). Both CD4$^+$ and CD8$^+$ maintain their natural auxiliary molecules to activate the TCR signal transduction pathway (Zhang and Wang 2019). A phase 1a/1b clinical study is also being conducted to determine the safety and tolerability of TCR therapy, which is cultured ex-vivo to target neoepitope (neoE), which gets presented exclusively on the surface of tumour cells (Cristea et al. 2021). Additionally, due to their tumour-restrictive expression pattern and escalated in-vivo immunogenicity, cancer testis antigens (CTA) are emerging as ideal targets for TCR mediated therapy. CTAs have also been observed to be important in early organogenesis and development (Wu and Münger 2021). The New York esophageal-1 (NY-ESO-1) and melanoma-associated antigen 4 (MAGE-$A_4$) are currently used as antigens of choice to target ovarian cancer, as they are predominantly expressed by the ovarian cancer cells (3.6% and 26.4% respectively) (Kerkar et al. 2016). In an ongoing phase I multi-tumour study (NCT03132922), MAGE-A4$^{c1032}$T cells are being administered to patients with MAGE-$A_4$ positive tumours, which also include ovarian cancer, to evaluate and assess the response. Currently, a panel of clinical trials (NCT03017131, NCT03691376, NCT02650986) are underway that involve the use of engineered T-cells to target NY-ESO-1 in ovarian cancer patients, but the results are yet to be published. Mesothelin, which is a glycosylphosphatidylinositol anchored cell surface protein observed to be highly expressed in ovarian cancer cells, helps in the peritoneal metastasis via its interaction with CA125. Moreover, the high expression of mesothelin has also been linked to its ability to confer pro-survival and proliferative advantages to the cancer cells, alongside promoting chemoresistance (Tang et al. 2013). Anderson et al. monitored the therapeutic efficacy of engineered CD8$^+$ T-cells ($TCR_{1045}$) to target ovarian cancer cells both in-vitro and in-vivo. These T cells showed promising results in $ID8_{VEGF}$ ovarian cancer cells. However, in murine models, $TCR_{1045}$ showed compromised persistence, which was also accompanied with the inability of the cells to produce cytokines by day 21. Treatment with successively sustained doses of T-cells maintained their anti-tumoural activity and prolonged their survival (Anderson et al. 2019).

### 4.6.6 Targeting the Tumour Microenvironment

The tumour microenvironment (TME) plays an important role in fostering pro-tumourigenic factors that aid in tumour progression and maintenance in the long run. These pro-tumourigenic factors involve both cellular and non-cellular components such as the extracellular matrix (ECM), stromal cells (mesenchymal stromal cells, fibroblasts, pericytes, adipocytes, blood, and lymphatic vascular networks), as well as immune cells (B and T lymphocytes, tumour-associated macrophages, and natural killer cells) (Wang et al. 2017). Reports have already highlighted the mechanisms through which each of these individual components contributes toward tumour growth. Therefore, therapeutic strategies targeting TME are also being studied to enhance therapy response. Among the non-cellular components, hypoxia is responsible for triggering a series of cellular responses so as to counteract detrimental effects of anoxic environment on the cells. This is primarily achieved by enhancing the stability of hypoxia-induced factor-1 (HIF-1) (Weidemann and Johnson 2008). It is a dimeric transcription factor that is composed of two subunits mainly, HIF-1α and HIF-1β. Both these subunits get expressed constitutively in the cells but under normoxic conditions HIF-1α undergoes proteasomal degradation. This degradation is mediated by prolyl-hydroxylase enzyme-2 (PHD-2) which hydroxylates the oxygen dependent-degradation domain of HIF-1α, thereby priming it for further ubiquitylation (Ziello et al. 2007). Under hypoxic conditions, HIF-1 is responsible in transcribing a panel of genes responsible for glucose metabolism, cell proliferation, macrophage polarization, angiogenesis, and metastasis (Roma-Rodrigues et al. 2019). Vascular endothelial growth factor (VEGF) is one of the most important downstream targets responsible for tumour angiogenesis, which gets upregulated by HIF-1 during hypoxic environment. The neo-vascularization of the tumour allows adequate blood flow to the cells thereby preventing the cytotoxic effects of hypoxia. Bevacizumab is the only anti-angiogenic agent that has been approved for the treatment of ovarian cancer. However, multiple small molecules are being repeatedly tried and tested for their efficacy as a promising alternative. Cediranib, developed by Aztrazeneca, is an oral inhibitor with multiple targets, including all three VEGFRs (VEGF-1, VEGF-2, and VEGFR3), c-KIT, and platelet-derived growth factor receptor alpha and beta. It is a potent ATP-competitive inhibitor in VEGF signaling cascade (Orbegoso et al. 2017). Preclinical studies indicated the ability of cediranib to significantly reduce progression and improve the survival of patient-derived ovarian cancer xenografts (Decio et al. 2015). Albeit promising results obtained in phase I and phase II trials involving the combination of cediranib and olaparib, the phase III trial reported that the combinatorial regime demonstrated a similar activity to the standard of care in relapsed platinum-sensitive patients (Liu et al. 2022). Cediranib is still being studied in combination with olaparib in advanced ovarian cancer in multiple clinical trials including Phase II/III GY005 trial (sponsored by the NCI) and Phase III ICON9 trial (sponsored by University College, London).

Chiauranib is another novel inhibitor that inhibits multiple angiogenesis-related kinases, chronic inflammation-related kinase CSF-1R, and mitosis-related kinase Aurora B. The phase III CHIPRO trial is a randomized, double-blind, 2-arm study to evaluate safety and efficacy profiles of the combination of chiauranib along with paclitaxel in platinum-refractory or platinum-resistant EOC patients (NCT04921527).

In the cellular forefront, macrophages play a crucial role in immune-modulation, which drives tumourigenesis. Derived from monocytes, macrophages get recruited into TME by both tumour cell-derived and adjacent stromal cell-derived chemokines. Depending upon the type of chemokine, monocytes differentiate into two subsets of macrophages, namely the M1-subtype, which gets "classically activated" by interferon gamma (IFN-$\gamma$); and the M2-subtype that are "alternatively activated" upon exposure to a range of diverse cytokines such as interleukin 4 (IL-4), IL-6, IL-10, TGF-β, macrophage colony stimulating factor (M-CSF), etc. (Roma-Rodrigues et al. 2019). The M2-subtype is notoriously known as tumour-associated macrophages (TAM) and is responsible for not only providing support to the growing neoplasm but also shielding the tumours from triggering a potent immune response by suppressing the system. Colony stimulating factor 1 (CSF1) or M-CSF is a secretory cytokine that binds onto its cognate receptor (CSF-1R) present in hematopoietic stem cells (HSC) allowing them to differentiate into macrophages (Boulakirba et al. 2018). Pfizer has developed a neutralizing G2 monoclonal antibody, PD-0360324, which specifically binds to CSF1 and prevents ligand-receptor interaction therefore inhibiting the relay of signals necessary for committing the HSCs to undergo macrophage differentiation. A phase II trial is being conducted to study the effect of a combinatorial administration of PD-0360324 and cyclophosphamide in treating patients with HGSOC (NCT02948101). Additionally, multiple small molecules and monoclonal antibodies against CSF1R are also in different phases of clinical trials for ovarian cancer such as PLX3397 (NCT02452424), Emactuzumab (NCT02323191), and Cabiralizumab (NCT02526017).

### 4.6.7 Vaccine-Based Therapies

#### 4.6.7.1 Cell-Based Vaccine—Cvac

The concept of vaccine development against cancer employs the ability of the body's humoral and cellular immune responses by presenting the cells with tumour specific antigens. This exposure allows the body to mount an anti-tumour response during tumourigenesis (Chow et al. 2020). Dendritic cells (DCs) have emerged as an excellent candidate for the development of vaccines especially due to their involvement in both the innate and adaptive immune responses. They are primarily antigen-presenting cells that are responsible for capturing, processing, and priming the body's immune system to develop an immune response upon further antigenic attack (Sabado et al. 2017). Developing DC-based vaccines majorly involves peptide-loaded or tumour lysate-loaded DCs as major approaches. In case of peptide-loaded DCs, the DCs are introduced to one or more recombinant antigenic peptides/proteins specific to the tumour prior to reinfusion, and these primed DCs are then injected back into the patient (Chow et al. 2020). To that effect, a variety of different dendritic cell vaccines have been developed against ovarian cancer, which involves proto-oncogene HER-2/neu-derived peptides such as GP2 (transmembrane part containing amino acids 654–662) and E75 (epitope recognized by cytotoxic T lymphocytes, amino acids 369–377) (Brossart et al. 2000). Lapuleucel-T is an investigational immunotherapeutic agent involving the exposure of autologous DCs to recombinant fusion antigen BA7072, which consists of both intra- and extra-cellular domains linked to granulocyte-macrophage colony-stimulating factor. Peethambaram et al. evaluated the safety, efficacy, and immunological activity of lapuleucel-T in HER-2/neu-expressing cancer patients. Immunological responses were observed in both colorectal and ovarian cancer patients (Peethambaram et al. 2009). Epithelial mucin 1 (MUC1) is a heavily glycosylated single pass type I transmembrane protein found to be expressed in 90% of ovarian cancer samples (Nath and Mukherjee 2014). Brossart et al. have also studied the effect of DC vaccines pulsed with MUC1-derived M1.1 and M1.2 9-mer peptides (Brossart et al. 2000). In 5/10 patients that were vaccinated, peptide-specific $CD8^+$ T cells were observed after 3 vaccinations in the peripheral blood. Interestingly, 2 patients showed the presence of (i) CAP-1 (CEA) and M3-271

(MAGE-3) peptide-specific T cells upon immunization with MUC1 peptides and (ii) MUC1 peptide-specific CTLs upon HER-2/neu-derived immunization. This cross-priming could be because of the destruction of tumour cells due to multiple rounds of vaccination, which may lead to tumour cell uptake and processing by DCs (Brossart et al. 2000). Rahma et al., used wt p53 peptide 264-272 pulsed DCs in their pilot phase II trial to test the efficacy of two different routes of administration: subcutaneous (SC) and intravenous (IV). Both SC and IV administration generated remarkable immunogenic responses: 69% and 83% respectively. Furthermore, there were no significant differences between these routes in their median overall survival and median progression free survival (Rahma et al. 2012). Promising results were obtained in the CAN-003 trial, which evaluated the efficacy of CVac-DC vaccine targeting MUC-1 glycoprotein in advanced EOC patients, as maintenance therapy. Mannan coupled with FP, a recombinant human fusion protein (mucin 1-glutathione S-transferase) acted as the antigen for CVac (Gray et al. 2016). For an improved immunogenic and clinical response rate, DCs exposed to whole tumour cell lysates obtained from ovarian cancer cell lines, fresh tumour biopsy samples and/or paraffin blocks of allogenic tumour samples have also been utilized for vaccine development (Zhang et al. 2020). To that effect, Tanyi et al. conducted a pilot clinical trial to test the ability of DCs pulsed with oxidized autologous whole-tumour cell lysates (OCDC) to illicit T cell responses in immunotherapy-naïve, platinum-treated recurrent ovarian cancer patients. The group observed a significant immune-reactivity to autologous tumour antigens, thereby conferring a prolonged survival. Additionally, the T cells also responded to unrecognized neoepitopes derived from non-synonymous somatic mutations in the tumour (Tanyi et al. 2018).

Reports documenting the effect of mRNA transfected DCs are also emerging as a potential technique for the generation of cell-based vaccines (Zhang et al. 2020). OX40 ligand (OX40L) is a co-stimulatory membrane bourn protein which is responsible in activating T cells to mount an immune response against cancer cells. mRNA-2416 developed by ModernaTX encodes human OX40L encapsulated within a lipid nanoparticle and is currently studied as an interventional agent singly or in combination with durvalumab in ovarian cancer (NCT03323398). The OLIVIA trial is also underway in order to determine the ability of RNA-lipoplex (RNA-LPX) as an interventional therapeutic along with neo-adjuvant chemotherapy to induce anti-tumour immune response. RNA-LPX is a liposome formulated mRNA vaccine containing three different mRNAs which target one EOC tumour associated antigen (NCT04163094).

#### 4.6.7.2 p53 Vaccine—p53MVA

Mutations in p53 ubiquitously act as driver mutations in the majority of solid tumours especially in ovarian cancer. The presence of anti-p53 reactive T cells in patients is suggestive of an incomplete tolerance toward p53, which is currently being exploited in the development of vaccines (Song et al. 2011). Moreover, the concentration of normal p53 in healthy cells is significantly low in comparison to their mutant counterpart, due to which p53 acts as a lucrative antigen for immune cells in a wide range of malignancies. Unfortunately, a significant variability exists in the p53 mutational sites among tumour types and patients, due to which p53 based vaccines commonly use the wild-type (wt) sequence (Hardwick et al. 2018). The effectivity of targeting p53 epitopes was previously studied using DCs pulsed with an adenoviral construct expressing wt-p53 in small-cell lung cancer. However, the vaccination failed to illicit the desired result, as it was only effective in just over half of the patients and was unable to maintain antigen-specific immune responses (Antonia et al. 2006). Further, the use of adenovirus and canarypox viruses expressing recombinant p53 has been discouraging, as no substantial clinical response was observed (Song et al. 2007). Leffers et al., used overlapping 25-30 amino acid synthetic long peptides (SLP) representing the most immunogenic parts of wt-p53 protein. Due to their size, they are effectively taken up by antigen presenting cells (APCs), where they are processed and thereby presented on the surface. The p53-SLP was successfully able to induce $CD4^+$ Th2-mediated cellular responses in 90% of ovarian cancer patients (Leffers et al. 2012). In 2011, Song et al., studied the ability of recombinant modified vaccinia virus Ankara (MVA) expressing wt-p53 to induce an immunogenic response ex vivo in cancer patients. They observed that 14 out of 23 patients successfully generated p53-specific IFN-$\gamma$ releasing cytotoxic T cell responses. These p53 responsive CTLs undergo degranulation and bring about the lysis of p53 overexpressing tumour cells (Song et al. 2011). Hardwick et al. conducted a phase I trial of p53MVA in combination with immunomodulatory chemotherapy agent gemcitabine in platinum-resistant ovarian cancer patients showing tumoural overexpression of p53. Gemcitabine was given prior to p53MVA administration to enhance the vaccine response by suppressing the immune system. They observed elevated $CD4^+$ and $CD8^+$ T cells ($\geqq 0.5\%$ of $CD4^+$ or $CD8^+$ population; called p53-responders), the levels of which were further correlated with longer PFS in comparison to non-p53 responders (Hardwick et al. 2018). A phase II clinical trial is also studying the combinatorial effect of treating recurrent platinum-resistant ovarian cancer patients with p53MVA along with pembrolizumab (anti-PD-1) (NCT03113487).

#### 4.6.7.3 Tumour-Associated Antigens—DPX-Survivac

Among the well-known repertoire of tumour-associated antigens (TAA), survivin is emerging as an attractive target for T lymphocyte-mediated tumour immunotherapy. Survivin is a member of the inhibitor-of-apoptosis proteins (IAP), which is specifically expressed in tissues during fetal development and tumourigenesis but not in normal tissues. The inhibition of apoptosis via survivin is speculated to be through the dissociation of Apfa-1-caspase 9 complex. Due to which the executioner caspase 3 fails to relay downstream apoptotic signaling cascade (Tamm et al. 1998). Cohen et al. observed that survivin was expressed in 74% of ovarian carcinoma correlating to poor prognostic parameters (Cohen et al. 2003). ImmunoVaccine Technologies Inc. developed a novel cancer vaccine, DPX-Survivac, which is designed to target survivin. In 2011, the phase I trial was carried out to investigate the safety and toxicity profiles of DPX-Survivac upon co-administering it with low-dose cyclophosphamide to late-stage ovarian cancer patients. DPX-Survivac was safely tolerated by all the recruited patients, produced survivin-specific polyfunctional T-cell immune responses and persisted in the system for several months, establishing it as an ideal vaccine candidate to be used in combination with different immunotherapeutics (Berinstein et al. 2015). A phase II trial is also studying the efficacy of administering DPX-Survivac along with cyclophosphamide and pembrolizumab in patients with advanced and recurrent solid tumours, which includes ovarian cancer (NCT03836352).

## 4.7 Conclusion

Despite low incidence, epithelial ovarian cancer is still a challenge due to its inconspicuous growth, ill-defined origin and extreme heterogeneity both at molecular and histological level. Slow pace integration of new therapeutic strategies over last two decades and acquirement of resistance towards platinum and other drugs on one hand add more difficulty to successful treatment and on another create more scope for future therapies. The classical knowledge on intra and inter tumoural heterogeneity of molecular subtypes of EOC is now revisited with the new knowledge regarding genetic and epigenetic heterogeneity. New clinical trials conducted with new-generation drugs majorly comprise targeted therapeutics in monotherapy or combinatorial settings. Similarly, new molecular mechanisms altered and rewired pathways are being identified by genome and proteome analysis for chemo resistant EOCs which paves new therapeutic options and hope for the relapsed patients. A number of such strategies are detailed here. It is still difficult to decide which one/s would be the prime choice/s, but with precision medicinal approach, higher PFS and OS are anticipated for this deadly and highly chemoresistant cancer in the near future.

## References


Agarwal, R. and S.B. Kaye, *Ovarian cancer: strategies for overcoming resistance to chemotherapy.* Nat Rev Cancer, 2003. **3**(7): p. 502–16.

Alcázar, J.L., et al., *IOTA simple rules for discriminating between benign and malignant adnexal masses: prospective external validation.* Ultrasound Obstet Gynecol, 2013. **42**(4): p. 467–71.

Amaravadi, R.K., et al., *Principles and current strategies for targeting autophagy for cancer treatment.* Clinical cancer research, 2011. **17**(4): p. 654–666.

Amit, M.O., et al., *Efficacy and Safety of Bevacizumab-Containing Therapy in Newly Diagnosed Ovarian Cancer: ROSiA Single-Arm Phase 3B Study.* International Journal of Gynecologic Cancer, 2017. **27**(1): p. 50.

An, Y., et al., *Development of a novel autophagy-related prognostic signature for serous ovarian cancer.* Journal of Cancer, 2018. **9**(21): p. 4058.

Anderson, K.G., et al., *Engineered Adoptive T-cell Therapy Prolongs Survival in a Preclinical Model of Advanced-Stage Ovarian Cancer.* Cancer Immunol Res, 2019. **7**(9): p. 1412–1425.

André, F., et al., *Alpelisib for PIK3CA-Mutated, Hormone Receptor–Positive Advanced Breast Cancer.* New England Journal of Medicine, 2019. **380**(20): p. 1929–1940.

Antonia, S.J., et al., *Combination of p53 cancer vaccine with chemotherapy in patients with extensive stage small cell lung cancer.* Clin Cancer Res, 2006. **12**(3 Pt 1): p. 878–87.

Arora, S., et al., *FDA Approval Summary: Olaparib Monotherapy or in Combination with Bevacizumab for the Maintenance Treatment of Patients with Advanced Ovarian Cancer.* Oncologist, 2021. **26**(1): p. e164–e172.

Assem, H., et al., *High-grade Endometrioid Carcinoma of the Ovary: A Clinicopathologic Study of 30 Cases.* The American Journal of Surgical Pathology, 2018. **42**(4): p. 534–544.

Baekelandt, M., et al., *Clinical significance of apoptosis-related factors p53, Mdm2, and Bcl-2 in advanced ovarian cancer.* Journal of clinical oncology, 1999. **17**(7): p. 2061–2061.

Barber, L.J., et al., *Secondary mutations in BRCA2 associated with clinical resistance to a PARP inhibitor.* The Journal of pathology, 2013. **229**(3): p. 422–429.

Bast, R.C., Jr., B. Hennessy, and G.B. Mills, *The biology of ovarian cancer: new opportunities for translation.* Nat Rev Cancer, 2009. **9**(6): p. 415–28.

Begicevic, R.-R. and M. Falasca, *ABC transporters in cancer stem cells: beyond chemoresistance.* International journal of molecular sciences, 2017. **18**(11): p. 2362.

Bell, D.A. and R.E. Scully, *Early de novo ovarian carcinoma. A study of fourteen cases.* Cancer, 1994. **73**(7): p. 1859–64.

Bell, D., et al., *Integrated genomic analyses of ovarian carcinoma.* Nature, 2011. **474**(7353): p. 609–615.

Bellio, C., et al., *PARP Inhibition Induces Enrichment of DNA Repair–Proficient CD133 and CD117 Positive Ovarian Cancer Stem CellsOvarian CSCs Mediate Resistance to PARP Inhibition.* Molecular Cancer Research, 2019. **17**(2): p. 431–445.

Bergers, G. and D. Hanahan, *Modes of resistance to anti-angiogenic therapy.* Nature Reviews Cancer, 2008. **8**(8): p. 592–603.

Berinstein, N.L., et al., *Survivin-targeted immunotherapy drives robust polyfunctional T cell generation and differentiation in advanced ovarian cancer patients.* Oncoimmunology, 2015. **4**(8): p. e1026529.

Bindra, R.S., M.E. Crosby, and P.M. Glazer, *Regulation of DNA repair in hypoxic cancer cells.* Cancer Metastasis Rev, 2007. **26**(2): p. 249–60.

Binju, M., et al., *Therapeutic inducers of apoptosis in ovarian cancer.* Cancers, 2019. **11**(11): p. 1786.

Birrer, M.J., et al., *Is Targeting the Folate Receptor in Ovarian Cancer Coming of Age?* Oncologist, 2019. **24**(4): p. 425–429.

Bishnu, A., et al., *Molecular imaging of the kinetics of hyperactivated ERK1/2-mediated autophagy during acquirement of chemoresistance.* Cell death & disease, 2021. **12**(2): p. 1–18.

Bitler, B.G., et al., *ARID1A-mutated ovarian cancers depend on HDAC6 activity.* Nature cell biology, 2017. **19**(8): p. 962–973.

Boulakirba, S., et al., *IL-34 and CSF-1 display an equivalent macrophage differentiation ability but a different polarization potential.* Sci Rep, 2018. **8**(1): p. 256.

Boyerinas, B., et al., *Let-7 modulates acquired resistance of ovarian cancer to Taxanes via IMP-1-mediated stabilization of multidrug resistance 1.* International journal of cancer, 2012. **130**(8): p. 1787–1797.

Bram, E.E., et al., *Chemotherapeutic drug-induced ABCG2 promoter demethylation as a novel mechanism of acquired multidrug resistance.* Neoplasia, 2009. **11**(12): p. 1359–IN11.

Brossart, P., et al., *Induction of cytotoxic T-lymphocyte responses in vivo after vaccinations with peptide-pulsed dendritic cells.* Blood, 2000. **96**(9): p. 3102–3108.

Castroviejo-Bermejo, M., et al., *A RAD51 assay feasible in routine tumor samples calls PARP inhibitor response beyond BRCA mutation. EMBO Mol Med 2018; 10: e9172.* PUBMED.

Catanzaro, D., et al., *Inhibition of glucose-6-phosphate dehydrogenase sensitizes cisplatin-resistant cells to death.* Oncotarget, 2015. **6**(30): p. 30102.

Chandra, A., et al., *Ovarian cancer: Current status and strategies for improving therapeutic outcomes.* Cancer Med, 2019. **8**(16): p. 7018–7031.

Chen, J., et al., *Evaluation of characteristics of CD44+ CD117+ ovarian cancer stem cells in three dimensional basement membrane extract scaffold versus two dimensional monocultures*. BMC cell biology, 2013. **14**(1): p. 1–11.

Chen, Y., M.M. Bieber, and N.N. Teng, *Hedgehog signaling regulates drug sensitivity by targeting ABC transporters ABCB1 and ABCG2 in epithelial ovarian cancer*. Molecular carcinogenesis, 2014. **53**(8): p. 625–634.

Chen, S., et al., *A Review of the Clinical Characteristics and Novel Molecular Subtypes of Endometrioid Ovarian Cancer*. Frontiers in Oncology, 2021. **11**.

Choi, Y.-J., et al., *Identifying an ovarian cancer cell hierarchy regulated by bone morphogenetic protein 2*. Proceedings of the National Academy of Sciences, 2015. **112**(50): p. E6882–E6888.

Chornokur, G., et al., *Common genetic variation in cellular transport genes and epithelial ovarian cancer (EOC) risk*. PLoS One, 2015. **10**(6): p. e0128106.

Chow, S., J.S. Berek, and O. Dorigo, *Development of Therapeutic Vaccines for Ovarian Cancer*. Vaccines (Basel), 2020. **8**(4).

Christie, E.L., et al., *Reversion of BRCA1/2 Germline Mutations Detected in Circulating Tumor DNA From Patients With High-Grade Serous Ovarian Cancer*. Journal of Clinical Oncology, 2017. **35**(12): p. 1274–1280.

Christie, E.L., et al., *Multiple ABCB1 transcriptional fusions in drug resistant high-grade serous ovarian and breast cancer*. Nature Communications, 2019. **10**(1): p. 1295.

Clamp, A. and G.C. Jayson, *Weekly platinum chemotherapy for recurrent ovarian cancer*. Br J Cancer, 2002. **86**(1): p. 2–4.

Cohen, C., et al., *Survivin Expression in Ovarian Carcinoma: Correlation with Apoptotic Markers and Prognosis*. Modern Pathology, 2003. **16**(6): p. 574–583.

Cragun, J.M., *Screening for ovarian cancer*. Cancer Control, 2011. **18**(1): p. 16–21.

Cree, I.A. and P. Charlton, *Molecular chess? Hallmarks of anti-cancer drug resistance*. BMC Cancer, 2017. **17**(1): p. 10.

Cristea, M., et al., *Abstract 331: Characterization of neoepitope (neoE)-specific T cells from peripheral blood for adoptive neoTCR-T cell therapy for patients with breast cancer (bc) or ovarian cancer (oc)*. Cancer Research, 2021. **81**(13_Supplement): p. 331–331.

Dai, F., Y. Zhang, and Y. Chen, *Involvement of miR-29b signaling in the sensitivity to chemotherapy in patients with ovarian carcinoma*. Human pathology, 2014. **45**(6): p. 1285–1293.

Decio, A., et al., *Cediranib combined with chemotherapy reduces tumor dissemination and prolongs the survival of mice bearing patient-derived ovarian cancer xenografts with different responsiveness to cisplatin*. Clin Exp Metastasis, 2015. **32**(7): p. 647–58.

Dev, H., et al., *Shieldin complex promotes DNA end-joining and counters homologous recombination in BRCA1-null cells*. Nature cell biology, 2018. **20**(8): p. 954–965.

Dinkic, C., et al., *Pazopanib (GW786034) and cyclophosphamide in patients with platinum-resistant, recurrent, pre-treated ovarian cancer - Results of the PACOVAR-trial*. Gynecol Oncol, 2017. **146**(2): p. 279–284.

DiSilvestro, P., et al., *Maintenance treatment of newly diagnosed advanced ovarian cancer: time for a paradigm shift?* Cancers, 2021. **13**(22): p. 5756.

Disis, M.L., et al., *Efficacy and Safety of Avelumab for Patients With Recurrent or Refractory Ovarian Cancer: Phase 1b Results From the JAVELIN Solid Tumor Trial*. JAMA Oncology, 2019. **5**(3): p. 393–401.

Do, K.T., et al., *Phase 1 Combination Study of the CHK1 Inhibitor Prexasertib and the PARP Inhibitor Olaparib in High-grade Serous Ovarian Cancer and Other Solid Tumors*. Clin Cancer Res, 2021. **27**(17): p. 4710–4716.

Dou, L. and Y. Zhang, *miR-4461 regulates the proliferation and metastasis of ovarian cancer cells and cisplatin resistance*. Frontiers in oncology, 2021. **11**.

Downie, D., et al., *Profiling cytochrome P450 expression in ovarian cancer: identification of prognostic markers*. Clinical Cancer Research, 2005. **11**(20): p. 7369–7375.

Elsnerova, K., et al., *Gene expression of membrane transporters: Importance for prognosis and progression of ovarian carcinoma*. Oncology reports, 2016. **35**(4): p. 2159–2170.

Emmanuel, C., et al., *Genomic classification of serous ovarian cancer with adjacent borderline differentiates RAS pathway and TP53-mutant tumors and identifies NRAS as an oncogenic driver*. Clin Cancer Res, 2014. **20**(24): p. 6618–30.

Farley, J., et al., *Selumetinib in women with recurrent low-grade serous carcinoma of the ovary or peritoneum: an open-label, single-arm, phase 2 study*. Lancet Oncol, 2013. **14**(2): p. 134–40.

Floquet, A., et al., *Progression-free survival by local investigator versus independent central review: comparative analysis of the AGO-OVAR16 Trial*. Gynecol Oncol, 2015. **136**(1): p. 37–42.

Fong, P.C., et al., *Poly(ADP)-ribose polymerase inhibition: frequent durable responses in BRCA carrier ovarian cancer correlating with platinum-free interval*. J Clin Oncol, 2010. **28**(15): p. 2512–9.

Fraser, M., et al., *Chemoresistance in human ovarian cancer: the role of apoptotic regulators*. Reproductive biology and endocrinology, 2003. **1**(1): p. 1–13.

Gadducci, A. and S. Cosio, *Therapeutic Approach to Low-Grade Serous Ovarian Carcinoma: State of Art and Perspectives of Clinical Research*. Cancers (Basel), 2020. **12**(5).

Gadducci, A., et al., *Clear cell carcinoma of the ovary: Epidemiology, pathological and biological features, treatment options and clinical outcomes*. Gynecol Oncol, 2021. **162**(3): p. 741–750.

Gao, J., et al., *TGF-β isoforms induce EMT independent migration of ovarian cancer cells*. Cancer cell international, 2014. **14**(1): p. 1–10.

Gao, N., et al., *Long non-coding RNAs: the regulatory mechanisms, research strategies, and future direc-*

*tions in cancers*. Frontiers in Oncology, 2020. **10**: p. 598817.

Giacomini, I., et al., *The pentose phosphate pathway and its involvement in cisplatin resistance*. International journal of molecular sciences, 2020. **21**(3): p. 937.

Gogola, E., et al., *Selective Loss of PARG Restores PARylation and Counteracts PARP Inhibitor-Mediated Synthetic Lethality*. Cancer Cell, 2018. **33**(6): p. 1078–1093.e12.

Gray, H.J., et al., *Progression-free and overall survival in ovarian cancer patients treated with CVac, a mucin 1 dendritic cell therapy in a randomized phase 2 trial*. Journal for ImmunoTherapy of Cancer, 2016. **4**(1): p. 34.

Gross, G., T. Waks, and Z. Eshhar, *Expression of immunoglobulin-T-cell receptor chimeric molecules as functional receptors with antibody-type specificity*. Proc Natl Acad Sci U S A, 1989. **86**(24): p. 10024–8.

Gulia, S., et al., *Pazopanib and Oral Cyclophosphamide in Women With Platinum-Resistant or -Refractory Epithelial Ovarian Cancer*. JCO Glob Oncol, 2020. **6**: p. 542–547.

Gulia, S., et al., *Maintenance therapy with a poly(ADP-ribose) polymerase inhibitor in patients with newly diagnosed advanced epithelial ovarian cancer: individual patient data and trial-level meta-analysis*. ESMO Open, 2022. **7**(5): p. 100558.

Gupta, S., et al., *Maintenance therapy for recurrent epithelial ovarian cancer: current therapies and future perspectives - a review*. J Ovarian Res, 2019. **12**(1): p. 103.

Haibe, Y., et al., *Resistance mechanisms to anti-angiogenic therapies in cancer*. Frontiers in oncology, 2020. **10**: p. 221.

Hajra, K.M., L. Tan, and J.R. Liu, *Defective apoptosis underlies chemoresistance in ovarian cancer*. Ovarian Cancer: State of the Art and Future Directions in Translational Research, 2008: p. 197–208.

Hamada, S.-I., et al., *Expression of glutathione S-transferase-π in human ovarian cancer as an indicator of resistance to chemotherapy*. Gynecologic oncology, 1994. **52**(3): p. 313–319.

Hamanishi, J., et al., *Safety and Antitumor Activity of Anti–PD-1 Antibody, Nivolumab, in Patients With Platinum-Resistant Ovarian Cancer*. Journal of Clinical Oncology, 2015. **33**(34): p. 4015–4022.

Hanahan, D. and R.A. Weinberg, *Hallmarks of cancer: the next generation*. Cell, 2011. **144**(5): p. 646–74.

Hardwick, N.R., et al., *p53-Reactive T Cells Are Associated with Clinical Benefit in Patients with Platinum-Resistant Epithelial Ovarian Cancer After Treatment with a p53 Vaccine and Gemcitabine Chemotherapy*. Clin Cancer Res, 2018. **24**(6): p. 1315–1325.

Hess, V., et al., *Mucinous Epithelial Ovarian Cancer: A Separate Entity Requiring Specific Treatment*. Journal of Clinical Oncology, 2004. **22**(6): p. 1040–1044.

Hollis, R.L., et al., *Molecular stratification of endometrioid ovarian carcinoma predicts clinical outcome*. Nature Communications, 2020. **11**(1): p. 4995.

Holzer, A.K., et al., *The copper influx transporter human copper transport protein 1 regulates the uptake of cisplatin in human ovarian carcinoma cells*. Molecular pharmacology, 2004. **66**(4): p. 817–823.

Hong, T., et al., *PARP inhibition promotes ferroptosis via repressing SLC7A11 and synergizes with ferroptosis inducers in BRCA-proficient ovarian cancer*. Redox Biology, 2021. **42**: p. 101928.

Hudson, C.D., et al., *Altered glutamine metabolism in platinum resistant ovarian cancer*. Oncotarget, 2016. **7**(27): p. 41637.

Iida, Y., et al., *Hypoxia promotes glycogen synthesis and accumulation in human ovarian clear cell carcinoma*. International journal of oncology, 2012. **40**(6): p. 2122–2130.

Jones, S., et al., *Frequent mutations of chromatin remodeling gene ARID1A in ovarian clear cell carcinoma*. Science, 2010. **330**(6001): p. 228–31.

Kakehi, M., et al., *Uridine diphosphate-glucuronosyltransferase (UGT) xenobiotic metabolizing activity and genetic evolution in pinniped species*. Toxicological Sciences, 2015. **147**(2): p. 360–369.

Kerkar, S.P., et al., *MAGE-A is More Highly Expressed Than NY-ESO-1 in a Systematic Immunohistochemical Analysis of 3668 Cases*. J Immunother, 2016. **39**(4): p. 181–7.

Kershaw, M.H., et al., *A phase I study on adoptive immunotherapy using gene-modified T cells for ovarian cancer*. Clin Cancer Res, 2006. **12**(20 Pt 1): p. 6106–15.

Kondrashova, O., et al., *Secondary somatic mutations restoring RAD51C and RAD51D associated with acquired resistance to the PARP inhibitor rucaparib in high-grade ovarian carcinoma*. Cancer discovery, 2017. **7**(9): p. 984–998.

Kondrashova, O., et al., *Methylation of all BRCA1 copies predicts response to the PARP inhibitor rucaparib in ovarian carcinoma*. Nature communications, 2018. **9**(1): p. 3970.

Koneru, M., et al., *A phase I clinical trial of adoptive T cell therapy using IL-12 secreting MUC-16(ecto) directed chimeric antigen receptors for recurrent ovarian cancer*. J Transl Med, 2015. **13**: p. 102.

Konstantinopoulos, P.A., et al., *Olaparib and α-specific PI3K inhibitor alpelisib for patients with epithelial ovarian cancer: a dose-escalation and dose-expansion phase 1b trial*. Lancet Oncol, 2019. **20**(4): p. 570–580.

Konstantinopoulos, P.A., et al., *A Phase 2 study of prexasertib (LY2606368) in platinum resistant or refractory recurrent ovarian cancer*. Gynecologic Oncology, 2022. **167**(2): p. 213–225.

Koshiyama, M., N. Matsumura, and I. Konishi, *Recent concepts of ovarian carcinogenesis: type I and type II*. Biomed Res Int, 2014. **2014**: p. 934261.

Kossaï, M., et al., *Ovarian Cancer: A Heterogeneous Disease*. Pathobiology, 2018. **85**(1-2): p. 41–49.

Kuhn, E., et al., *The pathogenesis of atypical proliferative Brenner tumor: an immunohistochemical and molecular genetic analysis*. Mod Pathol, 2014. **27**(2): p. 231–7.

Kuo, C.L., et al., *In vivo selection reveals autophagy promotes adaptation of metastatic ovarian cancer cells to abdominal microenvironment.* Cancer science, 2019. **110**(10): p. 3204–3214.

Kurman, R.J. and I.-M. Shih, *The Dualistic Model of Ovarian Carcinogenesis: Revisited, Revised, and Expanded.* The American Journal of Pathology, 2016. **186**(4): p. 733–747.

Kurzeder, C., et al., *Double-Blind, Placebo-Controlled, Randomized Phase III Trial Evaluating Pertuzumab Combined With Chemotherapy for Low Tumor Human Epidermal Growth Factor Receptor 3 mRNA-Expressing Platinum-Resistant Ovarian Cancer (PENELOPE).* J Clin Oncol, 2016. **34**(21): p. 2516–25.

Lan, C.-Y., et al., *Apatinib combined with oral etoposide in patients with platinum-resistant or platinum-refractory ovarian cancer (AEROC): a phase 2, single-arm, prospective study.* The Lancet Oncology, 2018. **19**(9): p. 1239–1246.

Larson, C.A., et al., *The role of the mammalian copper transporter 1 in the cellular accumulation of platinum-based drugs.* Molecular pharmacology, 2009. **75**(2): p. 324–330.

Lavarino, C., et al., *p53 gene status and response to platinum/paclitaxel-based chemotherapy in advanced ovarian carcinoma.* Journal of Clinical Oncology, 2000. **18**(23): p. 3936–3945.

Ledermann, J.A., et al., *Randomized phase II placebo-controlled trial of maintenance therapy using the oral triple angiokinase inhibitor BIBF 1120 after chemotherapy for relapsed ovarian cancer.* J Clin Oncol, 2011. **29**(28): p. 3798–804.

Ledermann, J.A., et al., *Cediranib in patients with relapsed platinum-sensitive ovarian cancer (ICON6): a randomised, double-blind, placebo-controlled phase 3 trial.* The Lancet, 2016. **387**(10023): p. 1066–1074.

Leffers, N., et al., *Long-term clinical and immunological effects of p53-SLP® vaccine in patients with ovarian cancer.* Int J Cancer, 2012. **130**(1): p. 105–12.

Leijen, S., et al., *Phase II Study of WEE1 Inhibitor AZD1775 Plus Carboplatin in Patients With TP53-Mutated Ovarian Cancer Refractory or Resistant to First-Line Therapy Within 3 Months.* Journal of Clinical Oncology, 2016. **34**(36): p. 4354–4361.

Lheureux, S., et al., *Adavosertib plus gemcitabine for platinum-resistant or platinum-refractory recurrent ovarian cancer: a double-blind, randomised, placebo-controlled, phase 2 trial.* Lancet, 2021. **397**(10271): p. 281–292

Li, T. and J. Wang, *Therapeutic effect of dual CAR-T targeting PDL1 and MUC16 antigens on ovarian cancer cells in mice.* BMC Cancer, 2020. **20**(1): p. 678.

Li, H., H. Xu, and H. Shen, *microRNA-106a modulates cisplatin sensitivity by targeting PDCD4 in human ovarian cancer cells.* Oncology letters, 2014. **7**(1): p. 183–188.

Li, J., et al., *Lipid desaturation is a metabolic marker and therapeutic target of ovarian cancer stem cells.* Cell stem cell, 2017. **20**(3): p. 303–314. e5.

Li, R., et al., *Sintilimab combined with bevacizumab in relapsed/persistent ovarian clear cell carcinoma (INOVA): an investigator-initiated, multicentre clinical trial-a study protocol of clinical trial.* BMJ Open, 2022. **12**(5): p. e058132.

Liang, Z.D., et al., *Mechanistic basis for overcoming platinum resistance using copper chelating agents.* Molecular cancer therapeutics, 2012. **11**(11): p. 2483–2494.

Lim, J., et al., *VEGFR3 Inhibition Chemosensitizes Ovarian Cancer Stemlike Cells through Down-Regulation of BRCA1 and BRCA2.* Neoplasia, 2014. **16**(4): p. 343–353.e2.

Lin, K.K., et al., *BRCA reversion mutations in circulating tumor DNA predict primary and acquired resistance to the PARP inhibitor rucaparib in high-grade ovarian carcinomaBRCA reversion mutations and resistance to PARP inhibitor.* Cancer discovery, 2019. **9**(2): p. 210–219.

Liontos, M., et al., *Profile of trebananib (AMG386) and its potential in the treatment of ovarian cancer.* Onco Targets Ther, 2014. **7**: p. 1837–45.

Lisio, M.A., et al., *High-Grade Serous Ovarian Cancer: Basic Sciences, Clinical and Therapeutic Standpoints.* Int J Mol Sci, 2019. **20**(4).

Liu, E., et al., *Overexpression of long non-coding RNA PVT1 in ovarian cancer cells promotes cisplatin resistance by regulating apoptotic pathways.* International journal of clinical and experimental medicine, 2015. **8**(11): p. 20565.

Liu, S., et al., *The effect of lncRNA HOTAIR on chemoresistance of ovarian cancer through regulation of HOXA7.* Biological Chemistry, 2018. **399**(5): p. 485–497.

Liu, J.F., et al., *Overall survival and updated progression-free survival outcomes in a randomized phase II study of combination cediranib and olaparib versus olaparib in relapsed platinum-sensitive ovarian cancer.* Annals of Oncology, 2019. **30**(4): p. 551–557.

Liu, J.F., et al., *Olaparib With or Without Cediranib Versus Platinum-Based Chemotherapy in Recurrent Platinum-Sensitive Ovarian Cancer (NRG-GY004): A Randomized, Open-Label, Phase III Trial.* J Clin Oncol, 2022. **40**(19): p. 2138–2147.

Lu, Z., et al., *The tumor suppressor gene ARHI regulates autophagy and tumor dormancy in human ovarian cancer cells.* The Journal of clinical investigation, 2008. **118**(12): p. 3917–3929.

Luvero, D., A. Milani, and J.A. Ledermann, *Treatment options in recurrent ovarian cancer: latest evidence and clinical potential.* Ther Adv Med Oncol, 2014. **6**(5): p. 229–39.

Marchetti, C., et al., *Chemotherapy resistance in epithelial ovarian cancer: Mechanisms and emerging treatments.* Semin Cancer Biol, 2021. **77**: p. 144–166.

Mathieu, K.B., et al., *Screening for ovarian cancer: imaging challenges and opportunities for improvement.* Ultrasound Obstet Gynecol, 2018. **51**(3): p. 293–303.

Matulonis, U.A., et al., *Ovarian cancer.* Nature Reviews Disease Primers, 2016. **2**(1): p. 16061.

McConechy, M.K., et al., *Ovarian and endometrial endometrioid carcinomas have distinct CTNNB1 and PTEN mutation profiles*. Mod Pathol, 2014. **27**(1): p. 128–34.

Mendel, D.B., et al., *In vivo antitumor activity of SU11248, a novel tyrosine kinase inhibitor targeting vascular endothelial growth factor and platelet-derived growth factor receptors: determination of a pharmacokinetic/pharmacodynamic relationship*. Clin Cancer Res, 2003. **9**(1): p. 327–37.

Miao, M., et al., *A phase II study of apatinib in patients with recurrent epithelial ovarian cancer*. Gynecologic Oncology, 2018. **148**(2): p. 286–290.

Milanesio, M.C., S. Giordano, and G. Valabrega, *Clinical implications of DNA repair defects in high-grade serous ovarian carcinomas*. Cancers, 2020. **12**(5): p. 1315.

Miliotou, A.N. and L.C. Papadopoulou, *CAR T-cell Therapy: A New Era in Cancer Immunotherapy*. Curr Pharm Biotechnol, 2018. **19**(1): p. 5–18.

Monk, B.J., et al., *Anti-angiopoietin therapy with trebananib for recurrent ovarian cancer (TRINOVA-1): a randomised, multicentre, double-blind, placebo-controlled phase 3 trial*. The Lancet Oncology, 2014. **15**(8): p. 799–808.

Monk, B.J., et al., *Final results of a phase 3 study of trebananib plus weekly paclitaxel in recurrent ovarian cancer (TRINOVA-1): Long-term survival, impact of ascites, and progression-free survival-2*. Gynecol Oncol, 2016. **143**(1): p. 27–34.

Moore, K.N., et al., *Adavosertib with Chemotherapy in Patients with Primary Platinum-Resistant Ovarian, Fallopian Tube, or Peritoneal Cancer: An Open-Label, Four-Arm, Phase II Study*. Clin Cancer Res, 2022. **28**(1): p. 36–44.

Mor, G., M.K. Montagna, and A.B. Alvero, *Modulation of apoptosis to reverse chemoresistance*. Apoptosis and Cancer: Methods and Protocols, 2008: p. 1–12.

Morales, J., et al., *Review of poly (ADP-ribose) polymerase (PARP) mechanisms of action and rationale for targeting in cancer and other diseases*. Crit Rev Eukaryot Gene Expr, 2014. **24**(1): p. 15–28.

Motzer, R.J., et al., *Activity of SU11248, a multitargeted inhibitor of vascular endothelial growth factor receptor and platelet-derived growth factor receptor, in patients with metastatic renal cell carcinoma*. J Clin Oncol, 2006. **24**(1): p. 16–24.

Nacson, J., et al., *BRCA1 mutation-specific responses to 53BP1 loss-induced homologous recombination and PARP inhibitor resistance*. Cell reports, 2018. **24**(13): p. 3513–3527. e7.

Nagaraj, A.B., et al., *Critical role of Wnt/β-catenin signaling in driving epithelial ovarian cancer platinum resistance*. Oncotarget, 2015. **6**(27): p. 23720.

Nasioudis, D., et al., *Undifferentiated carcinoma of the ovary: Epidemiology and prognosis of a rare tumor*. Gynecologic Oncology, 2017. **145**: p. 214.

Nath, S. and P. Mukherjee, *MUC1: a multifaceted oncoprotein with a key role in cancer progression*. Trends Mol Med, 2014. **20**(6): p. 332–42.

Niu, Y., et al., *A Novel Scoring System for Pivotal Autophagy-Related Genes Predicts Outcomes after Chemotherapy in Advanced Ovarian Cancer PatientsA Prognostic ATG-Based Scoring System in Ovarian Cancer*. Cancer Epidemiology, Biomarkers & Prevention, 2019. **28**(12): p. 2106–2114.

Nowicki, A., et al., *Ovarian Cancer and Cancer Stem Cells—Cellular and Molecular Characteristics, Signaling Pathways, and Usefulness as a Diagnostic Tool in Medicine and Oncology*. Cancers, 2021. **13**(16): p. 4178.

Obrist, F., et al., *Metabolic vulnerability of cisplatin-resistant cancers*. The EMBO journal, 2018. **37**(14): p. e98597.

Ogasawara, A., S. Sato, and K. Hasegawa, *Current and future strategies for treatment of ovarian clear cell carcinoma*. Journal of Obstetrics and Gynaecology Research, 2020. **46**(9): p. 1678–1689.

Ohishi, Y., et al., *ATP-binding cassette superfamily transporter gene expression in human primary ovarian carcinoma*. Clinical cancer research, 2002. **8**(12): p. 3767–3775.

Omar, H. and K.M. Rager, *Index*, in *Diagnosis and Management of Ovarian Disorders (Second Edition)*, A. Altchek, L. Deligdisch, and N.G. Kase, Editors. 2003, Academic Press: San Diego. p. 555–568.

Orbegoso, C., et al., *The role of Cediranib in ovarian cancer*. Expert Opin Pharmacother, 2017. **18**(15): p. 1637–1648.

Ovejero-Sánchez, M., R. González-Sarmiento, and A.B. Herrero, *Synergistic effect of Chloroquine and Panobinostat in ovarian cancer through induction of DNA damage and inhibition of DNA repair*. Neoplasia, 2021. **23**(5): p. 515–528.

Oza, A.M., et al., *A Biomarker-enriched, Randomized Phase II Trial of Adavosertib (AZD1775) Plus Paclitaxel and Carboplatin for Women with Platinum-sensitive TP53-mutant Ovarian Cancer*. Clin Cancer Res, 2020. **26**(18): p. 4767–4776.

Ozols, R.F., *Optimum chemotherapy for ovarian cancer*. International Journal of Gynecological Cancer, 2000. **10**(s1): p. 33–37.

Ozols, R.F., *Challenges for chemotherapy in ovarian cancer*. Ann Oncol, 2006. **17 Suppl 5**: p. v181–7.

Patch, A.-M., et al., *Whole–genome characterization of chemoresistant ovarian cancer*. Nature, 2015. **521**(7553): p. 489–494.

Peethambaram, P.P., et al., *A phase I trial of immunotherapy with lapuleucel-T (APC8024) in patients with refractory metastatic tumors that express HER-2/neu*. Clin Cancer Res, 2009. **15**(18): p. 5937–44.

Pelletier, J., et al., *Glycogen synthesis is induced in hypoxia by the hypoxia-inducible factor and promotes cancer cell survival*. Frontiers in oncology, 2012. **2**: p. 18.

Pettitt, S.J., et al., *Genome-wide and high-density CRISPR-Cas9 screens identify point mutations in PARP1 causing PARP inhibitor resistance*. Nature Communications, 2018. **9**(1): p. 1849.

Ponte, J.F., et al., *Mirvetuximab Soravtansine (IMGN853), a Folate Receptor Alpha-Targeting Antibody-Drug Conjugate, Potentiates the Activity of Standard of Care Therapeutics in Ovarian Cancer Models.* Neoplasia, 2016. **18**(12): p. 775–784.

Pujade-Lauraine, E., S. Banerjee, and S. Pignata, *Management of Platinum-Resistant, Relapsed Epithelial Ovarian Cancer and New Drug Perspectives.* J Clin Oncol, 2019. **37**(27): p. 2437–2448.

Radpour, R., *Tracing and targeting cancer stem cells: New venture for personalized molecular cancer therapy.* World journal of stem cells, 2017. **9**(10): p. 169.

Rahma, O.E., et al., *A gynecologic oncology group phase II trial of two p53 peptide vaccine approaches: subcutaneous injection and intravenous pulsed dendritic cells in high recurrence risk ovarian cancer patients.* Cancer Immunol Immunother, 2012. **61**(3): p. 373–84.

Rahma, O.E., et al., *Phase Ib study of pembrolizumab and trebananib (angiopoietin-2 inhibitor [Ang-2]): Preliminary analysis of the colorectal cancer (CRC) cohort.* Journal of Clinical Oncology, 2019. **37**(15_suppl): p. e14160–e14160.

Ray-Coquard, I., et al., *Final results from GCIG/ENGOT/AGO-OVAR 12, a randomised placebo-controlled phase III trial of nintedanib combined with chemotherapy for newly diagnosed advanced ovarian cancer.* Int J Cancer, 2020. **146**(2): p. 439–448.

Reles, A., et al., *Correlation of p53 mutations with resistance to platinum-based chemotherapy and shortened survival in ovarian cancer.* Clinical Cancer Research, 2001. **7**(10): p. 2984–2997.

Ricci, F., et al., *Overcoming platinum-acquired resistance in ovarian cancer patient-derived xenografts.* Therapeutic advances in medical oncology, 2019. **11**: p. 1758835919839543.

Roma-Rodrigues, C., et al., *Targeting Tumor Microenvironment for Cancer Therapy.* Int J Mol Sci, 2019. **20**(4).

Rose, M., et al., *PARP inhibitors: clinical relevance, mechanisms of action and tumor resistance.* Frontiers in cell and developmental biology, 2020. **8**: p. 564601.

Rottenberg, S., et al., *High sensitivity of BRCA1-deficient mammary tumors to the PARP inhibitor AZD2281 alone and in combination with platinum drugs.* Proceedings of the National Academy of Sciences, 2008. **105**(44): p. 17079–17084.

Sabado, R.L., S. Balan, and N. Bhardwaj, *Dendritic cell-based immunotherapy.* Cell Research, 2017. **27**(1): p. 74–95.

Sakai, W., et al., *Secondary mutations as a mechanism of cisplatin resistance in BRCA2-mutated cancers.* Nature, 2008. **451**(7182): p. 1116–1120.

Salazar, C., I.G. Campbell, and K.L. Gorringe, *When Is "Type I" Ovarian Cancer Not "Type I"? Indications of an Out-Dated Dichotomy.* Front Oncol, 2018. **8**: p. 654.

Sawers, L., et al., *Glutathione S-transferase P1 (GSTP1) directly influences platinum drug chemosensitivity in ovarian tumour cell lines.* British journal of cancer, 2014. **111**(6): p. 1150–1158.

Schmidt-Wolf, I.G., et al., *Use of a SCID mouse/human lymphoma model to evaluate cytokine-induced killer cells with potent antitumor cell activity.* J Exp Med, 1991. **174**(1): p. 139–49.

Sedláková, I., et al., *Clinical significance of the resistance proteins LRP, Pgp, MRP1, MRP3, and MRP5 in epithelial ovarian cancer.* International Journal of Gynecologic Cancer, 2015. **25**(2).

Seimetz, D., K. Heller, and J. Richter, *Approval of First CAR-Ts: Have we Solved all Hurdles for ATMPs?* Cell Med, 2019. **11**: p. 2155179018822781.

Shih Ie, M. and R.J. Kurman, *Ovarian tumorigenesis: a proposed model based on morphological and molecular genetic analysis.* Am J Pathol, 2004. **164**(5): p. 1511–8.

Shih, I.-M., Y. Wang, and T.-L. Wang, *The Origin of Ovarian Cancer Species and Precancerous Landscape.* The American Journal of Pathology, 2021. **191**(1): p. 26–39.

Son, H. and A. Moon, *Epithelial-mesenchymal transition and cell invasion.* Toxicological research, 2010. **26**: p. 245–252.

Song, G.-Y., et al., *An MVA vaccine overcomes tolerance to human p53 in mice and humans.* Cancer Immunology, Immunotherapy, 2007. **56**(8): p. 1193–1205.

Song, G.Y., et al., *Recombinant modified vaccinia virus ankara (MVA) expressing wild-type human p53 induces specific antitumor CTL expansion.* Cancer Invest, 2011. **29**(8): p. 501–10.

Soslow, R.A., *Histologic subtypes of ovarian carcinoma: an overview.* Int J Gynecol Pathol, 2008. **27**(2): p. 161–74.

Statello, L., et al., *Gene regulation by long non-coding RNAs and its biological functions.* Nature reviews Molecular cell biology, 2021. **22**(2): p. 96–118.

Stewart, J.M., et al., *Phenotypic heterogeneity and instability of human ovarian tumor-initiating cells.* Proceedings of the National Academy of Sciences, 2011. **108**(16): p. 6468–6473.

Sun, C., et al., *miR-9 regulation of BRCA1 and ovarian cancer sensitivity to cisplatin and PARP inhibition.* Journal of the National Cancer Institute, 2013. **105**(22): p. 1750–1758.

Sun, S., et al., *Prognostic value and implication for chemotherapy treatment of ABCB1 in epithelial ovarian cancer: a meta-analysis.* PloS one, 2016. **11**(11): p. e0166058.

Sung, H., et al., *Global Cancer Statistics 2020: GLOBOCAN Estimates of Incidence and Mortality Worldwide for 36 Cancers in 185 Countries.* CA Cancer J Clin, 2021. **71**(3): p. 209–49. https://doi.org/10.3322/caac.21660. Epub 2021 Feb 4. PMID: 33538338.

Szakács, G., et al., *Targeting multidrug resistance in cancer.* Nature reviews Drug discovery, 2006. **5**(3): p. 219–234.

Tamm, I., et al., *IAP-family protein survivin inhibits caspase activity and apoptosis induced by Fas (CD95), Bax, caspases, and anticancer drugs.* Cancer Res, 1998. **58**(23): p. 5315–20.

Tang, Z., M. Qian, and M. Ho, *The role of mesothelin in tumor progression and targeted therapy.* Anticancer Agents Med Chem, 2013. **13**(2): p. 276–80.

Tanyi, J.L., et al., *Personalized cancer vaccine effectively mobilizes antitumor T cell immunity in ovarian cancer.* Sci Transl Med, 2018. **10**(436).

Tao, M. and X. Wu, *The role of patient-derived ovarian cancer organoids in the study of PARP inhibitors sensitivity and resistance: from genomic analysis to functional testing.* Journal of Experimental & Clinical Cancer Research, 2021. **40**(1): p. 338.

Tao, Y., et al., *Clinicopathological and prognostic significance of cancer stem cell markers in ovarian cancer patients: evidence from 52 studies.* Cellular Physiology and Biochemistry, 2018. **46**(4): p. 1716–1726.

Taylor, J. and W.G. McCluggage, *Ovarian seromucinous carcinoma: report of a series of a newly categorized and uncommon neoplasm.* Am J Surg Pathol, 2015. **39**(7): p. 983–92.

Tentori, L., et al., *Poly(ADP-ribose) polymerase (PARP) inhibition or PARP-1 gene deletion reduces angiogenesis.* Eur J Cancer, 2007. **43**(14): p. 2124–33.

Tong, X., et al., *Expression levels of MRP1, GST-π, and GSK3β in ovarian cancer and the relationship with drug resistance and prognosis of patients.* Oncology letters, 2019. **18**(1): p. 22–28.

van der Gun, B.T., et al., *EpCAM in carcinogenesis: the good, the bad or the ugly.* Carcinogenesis, 2010. **31**(11): p. 1913–1921.

Vang, R., M. Shih Ie, and R.J. Kurman, *Ovarian low-grade and high-grade serous carcinoma: pathogenesis, clinicopathologic and molecular biologic features, and diagnostic problems.* Adv Anat Pathol, 2009. **16**(5): p. 267–82.

Vergote, I., et al., *AGO-OVAR 16: A phase III study to evaluate the efficacy and safety of pazopanib (PZ) monotherapy versus placebo in women who have not progressed after first line chemotherapy for epithelial ovarian, fallopian tube, or primary peritoneal cancer—Overall survival (OS) results.* Journal of Clinical Oncology, 2018. **36**(15_suppl): p. 5518–5518.

Vergote, I., et al., *Trebananib or placebo plus carboplatin and paclitaxel as first-line treatment for advanced ovarian cancer (TRINOVA-3/ENGOT-ov2/GOG-3001): a randomised, double-blind, phase 3 trial.* Lancet Oncol, 2019. **20**(6): p. 862–876.

Wang, J. and G.S. Wu, *Role of autophagy in cisplatin resistance in ovarian cancer cells.* Journal of Biological chemistry, 2014. **289**(24): p. 17163–17173.

Wang, Y., S.J. Decker, and J. Sebolt-Leopold, *Knockdown of Chk1, Wee1 and Myt1 by RNA interference abrogates G2 checkpoint and induces apoptosis.* Cancer Biol Ther, 2004. **3**(3): p. 305–13.

Wang, F., et al., *Involvement of SRPK1 in cisplatin resistance related to long non-coding RNA UCA1 in human ovarian cancer cells.* Neoplasma, 2015. **62**(3): p. 432–438.

Wang, M., et al., *Role of tumor microenvironment in tumorigenesis.* J Cancer, 2017. **8**(5): p. 761–773.

Wang, J.-y., A.-q. Lu, and L.-j. Chen, *LncRNAs in ovarian cancer.* Clinica Chimica Acta, 2019. **490**: p. 17–27.

Wang, A.J., et al., *A review of recent advances on single use of antibody-drug conjugates or combination with tumor immunology therapy for gynecologic cancer.* Front Pharmacol, 2022. **13**: p. 1093666.

Watanabe, N., M. Broome, and T. Hunter, *Regulation of the human WEE1Hu CDK tyrosine 15-kinase during the cell cycle.* Embo j, 1995. **14**(9): p. 1878–91.

Ween, M., et al., *The role of ABC transporters in ovarian cancer progression and chemoresistance.* Critical reviews in oncology/hematology, 2015. **96**(2): p. 220–256.

Wei, W., et al., *PARP-1 may be involved in angiogenesis in epithelial ovarian cancer.* Oncol Lett, 2016. **12**(6): p. 4561–4567.

Weidemann, A. and R.S. Johnson, *Biology of HIF-1alpha.* Cell Death Differ, 2008. **15**(4): p. 621–7.

Wilken, J.A., et al., *EGFR/HER-targeted therapeutics in ovarian cancer.* Future Med Chem, 2012. **4**(4): p. 447–69.

Willmott, L.J. and J.P. Fruehauf, *Targeted therapy in ovarian cancer.* J Oncol, 2010. **2010**: p. 740472.

Wu, S.C. and K. Münger, *Role and Clinical Utility of Cancer/Testis Antigens in Head and Neck Squamous Cell Carcinoma.* Cancers (Basel), 2021. **13**(22).

Wu, J., et al., *Antitumor effect of IL-12 gene-modified bone marrow mesenchymal stem cells combined with Fuzheng Yiliu decoction in an in vivo glioma nude mouse model.* J Transl Med, 2021. **19**(1): p. 143.

Xiao, L., et al., *YAP induces cisplatin resistance through activation of autophagy in human ovarian carcinoma cells.* OncoTargets and therapy, 2016: p. 1105–1114.

Xu, Q., Y. Tang, and X. Wang, *LncRNA EBIC promoted proliferation, metastasis and cisplatin resistance of ovarian cancer cells and predicted poor survival in ovarian cancer patients.* Eur Rev Med Pharmacol Sci, 2018a. **22**(14): p. 4440–4447.

Xu, D., et al., *Peripherally derived T regulatory and γδ T cells have opposing roles in the pathogenesis of intractable pediatric epilepsy.* J Exp Med, 2018b. **215**(4): p. 1169–1186.

Yamanoi, K., et al., *Acquisition of a side population fraction augments malignant phenotype in ovarian cancer.* Scientific reports, 2019. **9**(1): p. 14215.

Yang, C., et al., *Epigenetic silencing of miR-130b in ovarian cancer promotes the development of multidrug resistance by targeting colony-stimulating factor 1.* Gynecologic oncology, 2012a. **124**(2): p. 325–334.

Yang, L., et al., *Altered microRNA expression in cisplatin-resistant ovarian cancer cells and upregulation of miR-130a associated with MDR1/P-glycoprotein-mediated drug resistance.* Oncology reports, 2012b. **28**(2): p. 592–600.

Yang, L., et al., *Metabolic shifts toward glutamine regulate tumor growth, invasion and bioenergetics in ovarian cancer.* Molecular systems biology, 2014. **10**(5): p. 728.

Zervantonakis, I.K., et al., *Systems analysis of apoptotic priming in ovarian cancer identifies vulnerabilities and predictors of drug response*. Nature communications, 2017. **8**(1): p. 365.

Zhang, J. and L. Wang, *The Emerging World of TCR-T Cell Trials Against Cancer: A Systematic Review*. Technol Cancer Res Treat, 2019. **18**: p. 1533033819831068.

Zhang, S.-F., et al., *TXNDC17 promotes paclitaxel resistance via inducing autophagy in ovarian cancer*. Autophagy, 2015a. **11**(2): p. 225–238.

Zhang, L., et al., *2-Deoxy-d-Glucose Sensitizes Human Ovarian Cancer Cells to Cisplatin by Increasing ER Stress and Decreasing ATP Stores in Acidic Vesicles*. Journal of biochemical and molecular toxicology, 2015b. **29**(12): p. 572–578.

Zhang, C., D. Zhang, and J.-X. Cheng, *Coherent Raman scattering microscopy in biology and medicine*. Annual review of biomedical engineering, 2015c. **17**: p. 415–445.

Zhang, S., et al., *Preclinical characterization of Sintilimab, a fully human anti-PD-1 therapeutic monoclonal antibody for cancer*. Antibody Therapeutics, 2018. **1**(2): p. 65–73.

Zhang, X., et al., *Dendritic Cell Vaccines in Ovarian Cancer*. Front Immunol, 2020. **11**: p. 613773.

Zhou, Y., et al., *MicroRNA-449a reduces cell survival and enhances cisplatin-induced cytotoxicity via downregulation of NOTCH1 in ovarian cancer cells*. Tumor Biology, 2014. **35**: p. 12369–12378.

Zhu, Z., et al., *CYP1B1 enhances the resistance of epithelial ovarian cancer cells to paclitaxel in vivo and in vitro*. International journal of molecular medicine, 2015. **35**(2): p. 340–348.

Ziello, J.E., I.S. Jovin, and Y. Huang, *Hypoxia-Inducible Factor (HIF)-1 regulatory pathway and its potential for therapeutic intervention in malignancy and ischemia*. Yale J Biol Med, 2007. **80**(2): p. 51–60.

Zinzi, L., et al., *ABC transporters in CSCs membranes as a novel target for treating tumor relapse*. Frontiers in Pharmacology, 2014. **5**: p. 163.

# 5 The Role of the Human Microbiome in Epithelial Ovarian Cancer

Diane Mahoney

**Abstract**

Ovarian cancer is the fifth-leading cause of cancer deaths among women due to the absence of available screening methods to identify early disease. Thus, prevention and early disease detection investigations are of high priority, surrounding a critical window of opportunity to better understand important pathogenic mechanisms of disease progression. Microorganisms modulate molecular interactions in humans that can influence states of health and disease, including ovarian cancer. While the mechanisms of infectious microbial invasion that trigger the immune-inflammatory axis are well studied in cancer research, the complex interactions that promote the transition of noninfectious healthy microbes to pathobiont expansion are less understood. As traditional research has focused on the influences of infectious pathogens on ovarian cancer development and progression, the impact of noninfectious microbes has gained scientific attention. The objective of this chapter is to summarize current evidence on the role of microbiota in epithelial ovarian cancer throughout disease.

D. Mahoney (✉)
Franklin D. Gaines & Beverly J. Gaines Tipton Endowed Professor of Oncology Nursing, University of Kansas School of Nursing, Kansas City, KS, USA
e-mail: dmahoney@kumc.edu

**Keywords**

Epithelial ovarian cancer · Human microbiome · Microbiota · Microorganisms · Bacteria · Dysbiosis · Gut microbes · Vaginal microbes · Gastrointestinal tract · Lower reproductive tract · Upper reproductive tract

## 5.1 Introduction

In the United States annually, nearly 20,000 women are diagnosed with ovarian cancer, and over 13,000 women die of the disease, contributing to the 207,252 ovarian cancer deaths worldwide (Siegel et al. 2023; Sung et al. 2021). Unfortunately, these numbers are predicted to rise to 307,000 deaths by the year 2040 (Cabasag et al. 2022). Ovarian cancer is the fifth leading cause of cancer deaths among women due to the absence of available highly sensitive and specific screening methods to identify early disease (Siegel et al. 2023; Mathieu et al. 2018). The tumor biomarker CA125 has been used as the traditional primary ovarian cancer marker for the past several decades (Charkhchi et al. 2020). While CA125 has been a valuable indicator for evaluating prognosis and the efficacy of chemotherapy, it is not sufficiently reliable in disease screening (Zhang et al. 1875). Early disease remains undetectable with current screening methods, and women with ovarian cancer are



H. Schatten (ed.), *Cell and Molecular Biology of Ovarian Cancer*, Advances in Experimental Medicine and Biology 1452, https://doi.org/10.1007/978-3-031-58311-7_5

often asymptomatic in the early stage of the disease and are undiagnosed until a more advanced stage (Doubeni et al. 2016). Some women have reported vague abdominal symptoms for up to 2 years prior to being diagnosed with ovarian cancer, but the absence of complementary biologic measures to better evaluate symptoms presents a serious challenge for early disease detection (Mahoney and Pierce 2022). Ninety percent of all ovarian cancers are epithelial ovarian cancers, and the majority (60%) are high-grade serous tumors that metastasize into the peritoneal cavity and have the fewest established risk factors (Surveillance 2020; Yousefi et al. 2020; Society 2021). The high-grade serous ovarian carcinoma (HGSOC) epithelial subtype is characterized by bilateral ovarian involvement, aggressive behavior, late-stage diagnosis, and high mortality (Lisio et al. 2019). Corroborating evidence substantiates that HGSOC originates in the fallopian tubes as serous tubal intraepithelial carcinoma precursor lesions with the same p53 gene mutation signature as observed in the ovarian tumors (Labidi-Galy et al. 2017; Medeiros et al. 2006; Kyo et al. 2020; Soong et al. 2018). The latency between the formation of STIC lesions and the initiation of metastatic ovarian carcinoma is between 6 and 7 years (Labidi-Galy et al. 2017; Kyo et al. 2020). Thus, disease prevention and early detection investigations are of high priority surrounding this critical window of opportunity to better understand important pathogenic mechanisms of disease progression within the fallopian tube as well as identify molecular determinants of ovarian susceptibility. As researchers apply ongoing rigorous methodologies to investigate ovarian carcinogenesis in both cases of sporadic disease and those of genetic predisposition (*BRCA1/2* gene mutation carriers), the role of the human microbiome has generated increasing interest.

## 5.2 Human Microbiome

The launch of the National Institutes of Health's (NIH) Human Microbiome Project has fostered imaginative discoveries as scientists have utilized advances in high-throughput sequencing technology to identify and characterize the genomes of microorganisms that colonize in various body regions to determine their influence on health and disease (Group, N.H.W. 2009). The human microbiome encompasses 10–100 trillion symbiotic microbes (bacteria, viruses, fungi, protozoa, and bacteriophages) that reside within the body (Rodrigues Hoffmann et al. 2016; Ursell et al. 2012). However, the majority of the microorganisms are bacteria that are harbored primarily in the gastrointestinal (GI) tract or gut (Ursell et al. 2012; Ahlawat et al. 2021). Microorganisms modulate molecular interactions in humans that can influence states of health and disease including cancer (Ogunrinola et al. 2020). The regulation of homeostasis including immune response and metabolism within the host can impact or become impacted by microbial dysbiosis—broadly defined as any change to the microbial composition of resident commensal communities relative to the communities present in healthy populations (Zheng et al. 2020; Petersen and Round 2014). Dysbiosis is also characterized by the progression away from microbial homeostasis through pathobiont expansion, changes in diversity, or the reduction of beneficial microbes (see Fig. 5.1) (Petersen and Round 2014; Martinez et al. 2021). Under normal conditions, a symbiotic relationship of commensalism is established between the human host and residing microbial communities that support their adaptation to the unique features of each niche without being attacked by the host immunity in the absence of an infection (Ogunrinola et al. 2020). Compositional changes of bacterial microbes are suspected to play a role in the development of cancers (Castano-Rodriguez et al. 2017; Ramirez-Labrada et al. 2020; Gagniere et al. 2016; Choi et al. 2021; Eslami et al. 2020; Schmidt et al. 2014). Microbial invasion of infectious pathogens promotes inflammatory reactions that mediate immune cells as a defense mechanism, and chronic inflammation is a hallmark of cancer and thus a risk factor for epithelial ovarian cancer (Xie et al. 2017; Comen et al. 2018). Evidence suggests that chronic inflammation produces a pro-tumorigenic microenvironment through

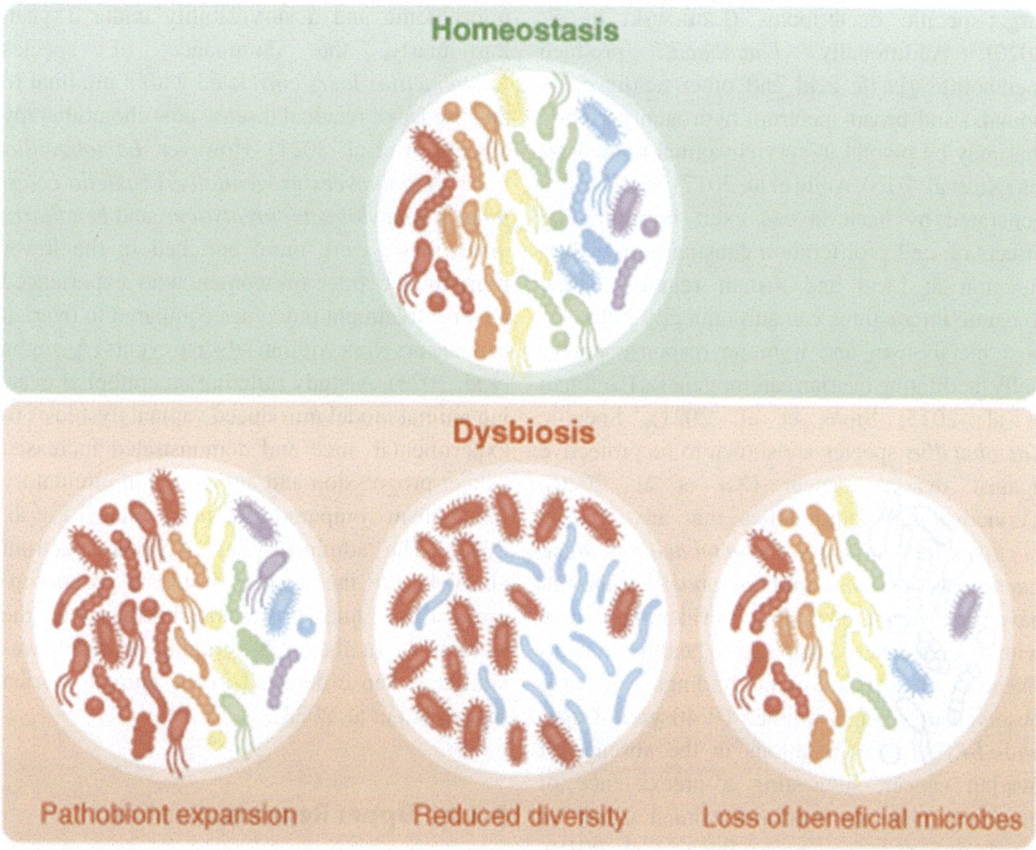


**Fig. 5.1** Microbial homeostasis and dysbiosis. (From Petersen and Round (2014). Permissions provided by John Wiley and Copyright Clearance Center January 23, 2024)

proinflammatory production, angiogenesis, and tissue remodeling (Comen et al. 2018). While the mechanisms of infectious microbial invasion that trigger the immune-inflammatory axis are well studied in cancer research, the complex interactions that promote the transition of noninfectious healthy microbes to pathobiont expansion are less understood.

## 5.3 Epithelial Ovarian Cancer and the Human Microbiome

### 5.3.1 Lower Reproductive Tract

The lower reproductive tract encompasses the vagina and cervix (Kumar et al. 2023). The vagina is a nonsterile environment dominated primarily by aerobic bacterial colonies (Smith and Ravel 2017). Consequently, vaginal microbial imbalances trigger immune responses, degrade the host mucosa, and increase vulnerability to the overgrowth of anaerobes causing bacterial vaginosis (BV) infection (Muzny et al. 2019). Scientists speculate that BV facilitates the migration of harmful microbes to the upper reproductive tract and has been associated with pelvic inflammatory disease (PID) (Chase et al. 2015; Piao et al. 2020). A healthy vaginal microbiome is considered a fundamental barrier to pathogenic invasion. *Lactobacillus* species play a valuable role in safeguarding the lower reproduction tract from harmful microorganisms by upregulating tight epithelial junction proteins that inhibit pathogen migration and improve tissue integrity (Xu et al. 2020). *Lactobacilli* also act as gatekeepers against microbial infections by maintaining low pH levels (<4.5) and producing target-specific bacteriocins (Laniewski et al. 2020). Additionally, *Lactobacilli* produce metabolites (lactic acid and other acidic compounds) and broad-spectrum hydrogen peroxide that may be pivotal to cervicovaginal homeostasis (Xu et al. 2020; Amin et al. 2017). Metabolites generated by bacteria can exert hormone-like effects on cell proliferation causing barrier dysfunction at local and distant regions. These dynamic interactions can subsequently influence host metabolism and immune response potentially facilitating ovarian carcinogenesis (Neuman et al. 2015; Sipos et al. 2021). Specific *Lactobacillus* species are shown to be protective against ovarian cancer (Xu et al. 2020). Cervicovaginal communities that are poor in *Lactobacillus crispatus, Lactobacillus iners, Lactobacillus gasseri, and Lactobacillus jensenii* have been shown to be more prevalent in women with ovarian cancer compared to healthy controls (Nene et al. 2019). Similar findings were also observed in younger women (< 40 yrs. of age) with *BRCA* gene mutations in the absence of ovarian cancer, suggesting a preexistence of microbial dysbiosis could put women at risk for developing ovarian cancer (Nene et al. 2019). When cervicovaginal microbial dysbiosis occurs, *Lactobacilli* have been replaced with mixed anaerobic bacteria (Laniewski et al. 2020). In comparison to the dominance of *Lactobacilli*-rich genera found in the cervicovaginal fluid of healthy controls, premenopausal women with epithelial ovarian cancer subtypes (serous, clear cell, and mucinous) have demonstrated a higher diversity of genera *Propionibacterium* and *Corynebacterium* (Morikawa et al. 2022). Thus, microbial dysbiosis could be highly favorable to the hallmarks of ovarian cancer through cervicovaginal pathobiont expansion at the time of chronic inflammation, barrier disruption, genomic instability, altered proliferation, apoptosis, and/or angiogenesis (Laniewski et al. 2020). Furthermore, distinct vaginal bacteria have been associated with chemotherapy resistance, disease recurrence, and survival. In a recent study, over one-third of women with epithelial ovarian cancer who had primary platinum-resistance demonstrated an *Escherichia*-dominated vaginal microbiome and a survivability under 1 year. Particularly, the dominance of species *Lactobacillus iners* correlated with a minimal to an absence of residual disease post chemotherapy (Jacobson et al. 2021). However, *Lactobacillus iners, Actinomyces urogenitalis, Ezakiella coagulans, Campylobacter ureolyticus,* and *Fenollaria massiliensis* were more enriched in the lower reproductive tract of women who experienced adverse treatment outcomes compared to ovarian cancer survivors without adverse events (Asangba et al. 2023). A study utilizing an epithelial ovarian animal model introduced vaginal dysbiosis to experimental mice and demonstrated increased cancer progression and alterations in circulatory metabolism compared to control mice (Li et al. 2023). The administration of broad-spectrum antibiotics to the vaginal dysbiosis group suppressed the tumor profession. Furthermore, the microbiota in the mouse vaginal samples was similar to those detected in the animal ovarian tumors (Li et al. 2023).

### 5.3.2 Upper Reproductive Tract

The upper reproductive tract encompasses the uterus, fallopian tubes, ovaries, and peritoneal fluid (Kumar et al. 2023). Fallopian tube inflammation is believed to contribute to ovarian cancer development, and PID has been associated with increased disease risk (Piao et al. 2020). Some investigators further hypothesize associations of infectious microbes including *Chlamydia trachomatis*, cytomegalovirus, and human papilloma virus with epithelial ovarian cancer while others have found no such relationship (Idahl et al. 2020; Shanmughapriya et al. 2012; Alibek et al. 2012; Fortner et al. 2019; Idahl et al. 2010; Ingerslev et al. 2019; Jonsson et al. 2018; Pathak et al. 2020; Svahn et al. 2014). In the absence of infection, microbial communities have been identified within the upper female reproductive tract, once assumed to be a sterile environment (Chen et al. 2017; Miles et al. 2017). Among women of reproductive age undergoing surgery for benign gynecologic conditions, replicate samples taken from the fallopian tubes, endometrium, and peritoneal fluid that have been consistently characterized by the same bacteria after different microbial gene sequencing runs. Specifically, bacteria present in peritoneal fluid were further validated with conventional culturing and isolation methods (Chen et al. 2017).

Recent advances are demonstrating relationships with bacterial interactions and ovarian tumor activity in cancer research (Ciernikova et al. 2022). Scientists have used high throughput microbial gene sequencing to compare the microbiota diversity and composition in ovarian tissues of women with and without disease followed by quantitative PCR validation methods in additional samples. Researchers found phylum *Proteobacteria* and species *Acinetobacter lwoffii* more in abundant epithelial ovarian cancer tumor tissue compared to normal distal fallopian tube tissues of women without disease, as well as differing transcriptional profiles (Zhou et al. 2019). Similarly, another study reported *Proteobacteria* and *Firmicutes* were the most dominant phyla detected in ovarian cancer tissue samples (Banerjee et al. 2017). One study reported bacteria of the upper reproductive tract of women with ovarian cancer to be of cervical origin yet distinct to those with disease (Qin et al. 2022). Others have reported microbial composition differences in ovarian tissue and peritoneal fluid between women with epithelial ovarian cancer and women with benign gynecologic conditions across numerous bacterial taxonomy (Miao et al. 2020; Wang et al. 2020). In addition, investigators reported increased distinct mechanistic pathways unique to ovarian cancer tissue samples compared to controls (Wang et al. 2020). Proximal fallopian tube, fimbriae, and ovarian tissue samples have revealed unique microbial profiles within each region in postmenopausal women with ovarian cancer and modest overall differences between those with and without disease (Brewster et al. 2022). Genera *Actinoplanes, Arthrobacter, Bradyrhizobium, Gemmatimonas, Limnobacter, Roseobacter, Saccharopolyspora, and Mycobacterium species* showed were differentially more abundance in the upper reproductive tract of women with ovarian cancer in comparison to controls. Two of eight subjects had a diagnosis of HGSOC (Brewster et al. 2022). Another study revealed higher bacterial diversity in peritoneal fluid and lower reproductive tract of women with early stage disease compared to late stage disease and benign conditions (Asangba et al. 2023). Similar differences were also observed when comparing bacterial diversity between HGSOC and low-grade serous ovarian carcinoma in the uterus and ovaries (Asangba et al. 2023). Because HGSOC originates in the mucosal epithelium of the fallopian tubes, bacterial characterization combined with mechanistic pathways in this region has prompted ongoing investigation to elucidate the ovarian tumor microenvironment.

### 5.3.3 Gastrointestinal Tract

The human GI tract begins with the oral cavity, continues through the stomach, intestines, and ends at the anus (Hillman et al. 2017). The majority of the human microbiome is harbored in the gut as trillions of bacteria (Ursell et al. 2012; Ahlawat et al. 2021). Gut microbial dysbiosis fosters epithelial barrier dysfunction and immune dysregulation that is tumor microenvironment permissive (D'Antonio et al. 2022). Gut microbes can reach the distal organs via the bloodstream, however, our understanding remains limited concerning gut microbiota-dependent mechanistic changes in ovarian carcinogenesis (Clarke et al. 2014; Obrenovich 2018). The contributions of gut dysbiosis have prompted recent investigations, provided that GI-type symptoms (i.e., abdominal pain, bloating, indigestion, and early satiety) are most commonly reported at diagnosis by women with ovarian cancer and persist throughout treatment (Mahoney and Pierce 2022). One study demonstrated that gut microbial diversity differentiated early- versus late-stage disease in women with epithelial ovarian cancer (Asangba et al. 2023). Animal models have shown that antibiotic induced gut microbial dysbiosis promotes tumor growth and causes an epithelial-mesenchymal transition (Xu et al. 2019). Ovarian tumor size and weight increased significantly faster among intestinal microbiota dysbiosis mice compared to control mice, yet this situation resolved with macrophage depletion. Additionally, macrophages collected from intestinal microbiota dysbiosis mice were more likely to promote the production of inflammatory cytokines tumor necrosis factor alpha and interleukin 6 (IL-6) (Xu et al. 2019). While these findings were paralleled in different studies, no significant alterations were observed in ovarian tumor growth when comparing gut dysbiosis mice to controls (Mahoney et al. 2022). In a separate mouse model study, TLR5 signaling at regions of bacterial colonization in ovarian tumor-bearing hosts initiated tumor-promoting systemic inflammation and the deployment of myeloid derived suppressor cells and immunosuppressive gamma delta T cells. Moreover, tumors that induced TLR5-dependent systemic IL-6 up-regulation responded to exogenous IL-6 by producing additional levels (Rutkowski et al. 2015).

Researchers have conducted human and animal studies to examine the role of gut microbial dysbiosis during ovarian cancer treatment. Fecal microbiota profiles were assessed on women who were newly diagnosed with epithelial ovarian cancer and throughout neoadjuvant and adjuvant chemotherapy. When compared to healthy women, gut microbial dysbiosis was evident in women with disease and platinum-resistant women exhibited reduced diversity of bacteria with increased proportions of genus *Bifidobacterium* and family *Coriobacteriaceae* (D'Amico et al. 2021). Moreover, women with epithelial ovarian cancer have shown reductions in the fecal abundance of phyla *Bacteroidetes* and *Firmicutes*, and an increased abundance in phylum *Proteobacteria* postoperatively compared to preoperatively. However, these observations were found to be opposite when comparing pre-chemotherapy to post-chemotherapy status (Tong et al. 2020). Bacterial species (*Blautia pseudococcoides*, *Bacteroides ovatus*, *Butyricicoccus faecihominis, Diasliter invisus, Lactobacillus gasseri*, *Sutterella wadsworthensis,* and *Veillonella nakazawae*) have been differentially enriched in the fecal samples of women who experienced treatment-related adverse events and those who were deceased compared to survivors who did not experience adverse events 2 years post-diagnosis (Asangba et al. 2023). Inversely, restoration of deficient bacteria has been shown to resolve microbial dysbiosis and provide beneficial effects. For example, fecal microbiota transplants repaired cisplatin-induced intestinal mucosal damage and promoted a reduction in serum IL-6 levels in ovarian tumor-bearing mice (Perales-Puchalt et al. 2018). Cecal microbiota transplants have reversed cisplatin resistance, restored circulating gut-derived metabolites, and prolonged survival of ovarian tumor bearing mice treated with broad-spectrum antibiotics (Chambers et al. 2022). To determine the influences of gut microbial dysbiosis in the most aggressive epithelial ovarian cancer subtype, investigators tested a HGSOC mouse model with the administration of antibiotics to the treatment group. In contrast to existing studies, mice with gut microbial dysbiosis had a reduction in advanced tumors compared to control mice (Chen et al. 2021). Moreover, the treatment of ovarian cancer cell lines with the genus *Bacillus* bacteria strains has induced significant inhibitory effects on the proliferation (Zhao et al. 2020).

## 5.4 Conclusions and Future Directions

This chapter has highlighted human microbes associated with epithelial ovarian cancer. Some investigators have described highly distinct differences in women with disease when compared to controls, whereas others have observed marginal differences. While certain microbiota are shown to be unique to women with disease, the multifaceted contributors to microbial-host homeostasis do not exist within a vacuum. Dietary preferences (consumption of high-fat diets, low-fiber, processed foods, artificial ingredients), lifestyle habits (sedentary activity, cigarette smoking, excessive alcohol intake), vaginal infections, stress, medication, and the envir onmental factors together contribute to microbiota dysbiosis (Ahlawat et al. 2021; Martinez et al. 2021; Redondo-Useros et al. 2020; Ahn and Hayes 2021; Lev-Sagie et al. 2022). For instance, high-fat diets are often accompanied by a lower intake of carbohydrates and fiber, which limits the accessibility of fermentation substrates for healthy bacteria to thrive (Redondo-Useros et al. 2020; Valdes et al. 2018). Hence, a woman's eating habits, in conjunction with the presence of *BRCA* gene mutations and other exposures, could influence the likelihood of gut bacteria dysbiosis and ovarian carcinogenesis. Future investigations are recommended to establish strong causal relationships that can directly link molecular mechanisms to clinical outcomes throughout the ovarian cancer continuum. Furthermore, improved study inclusion of racial and ethnically diverse women residing in varying sociodemographic regions is needed in microbiome research to ensure accurate interpretation and generalizability of the findings. Importantly, the establishment of robust microbial candidates that parallel with tumor biomarker CA125 levels would be a valuable step for the validation of microbial-based diagnostic and prognostic disease biomarkers. Such novel discoveries could also steer development of microbial-based preventative and therapeutic interventions tailored to women with epithelial ovarian cancer and those at high risk for disease.

## References


Ahlawat, S., Asha & Sharma, K.K. Gut-organ axis: a microbial outreach and networking. *Lett Appl Microbiol* **72**, 636–668 (2021).

Ahn, J. & Hayes, R.B. Environmental Influences on the Human Microbiome and Implications for Noncommunicable Disease. *Annu Rev Public Health* **42**, 277–292 (2021).

Alibek, K., Karatayeva, N. & Bekniyazov, I. The role of infectious agents in urogenital cancers. *Infectious Agents and Cancer* **7**, 35 (2012).

Amin, M., Moradi Choghakabodi, P., Alhassan Hamidi, M., Najafian, M. & Farajzadeh Sheikh, A. In vitro antimicrobial activities of metabolites from vaginal Lactobacillus strains against Clostridium perfringens isolated from a woman's vagina. *J Chin Med Assoc* **80**, 29–33 (2017).

Asangba, A.E., *et al.* Diagnostic and prognostic potential of the microbiome in ovarian cancer treatment response. *Sci Rep* **13**, 730 (2023).

Banerjee, S., *et al.* The ovarian cancer oncobiome. *Oncotarget* **8**, 36225–36245 (2017).

Brewster, W.R., *et al.* An evaluation of the microbiota of the upper reproductive tract of women with and without epithelial ovarian cancer. *Gynecol Oncol Rep* **42**, 101017 (2022).

Cabasag, C.J., *et al.* Ovarian cancer today and tomorrow: A global assessment by world region and Human Development Index using GLOBOCAN 2020. *Int J Cancer* **151**, 1535–1541 (2022).

Castano-Rodriguez, N., Goh, K.L., Fock, K.M., Mitchell, H.M. & Kaakoush, N.O. Dysbiosis of the microbiome in gastric carcinogenesis. *Sci Rep* **7**, 15957 (2017).

Chambers, L.M., *et al.* Disruption of the Gut Microbiota Confers Cisplatin Resistance in Epithelial Ovarian Cancer. *Cancer Res* **82**, 4654–4669 (2022).

Charkhchi, P., *et al.* CA125 and Ovarian Cancer: A Comprehensive Review. *Cancers (Basel)* **12**(2020).

Chase, D., Goulder, A., Zenhausern, F., Monk, B. & Herbst-Kralovetz, M. The vaginal and gastrointestinal microbiomes in gynecologic cancers: a review of applications in etiology, symptoms and treatment. *Gynecol Oncol* **138**, 190–200 (2015).

Chen, C., *et al.* The microbiota continuum along the female reproductive tract and its relation to uterine-related diseases. *Nat Commun* **8**, 875 (2017).

Chen, L., *et al.* Altering the Microbiome Inhibits Tumorigenesis in a Mouse Model of Oviductal High-Grade Serous Carcinoma. *Cancer Res* **81**, 3309–3318 (2021).

Choi, S., Chung, J., Cho, M.L., Park, D. & Choi, S.S. Analysis of changes in microbiome compositions related to the prognosis of colorectal cancer patients based on tissue-derived 16S rRNA sequences. *J Transl Med* **19**, 485 (2021).

Ciernikova, S., Sevcikova, A., Stevurkova, V. & Mego, M. Tumor microbiome - an integral part of the tumor microenvironment. *Front Oncol* **12**, 1063100 (2022).

Clarke, G., *et al.* Minireview: Gut microbiota: the neglected endocrine organ. *Mol Endocrinol* **28**, 1221–1238 (2014).

Comen, E.A., Bowman, R.L. & Kleppe, M. Underlying Causes and Therapeutic Targeting of the Inflammatory Tumor Microenvironment. *Front Cell Dev Biol* **6**, 56 (2018).

D'Amico, F., *et al.* Gut Microbiota Dynamics during Chemotherapy in Epithelial Ovarian Cancer Patients Are Related to Therapeutic Outcome. *Cancers (Basel)* **13**(2021).

D'Antonio, D.L., Marchetti, S., Pignatelli, P., Piattelli, A. & Curia, M.C. The Oncobiome in Gastroenteric and Genitourinary Cancers. *Int J Mol Sci* **23**(2022).

Doubeni, C.A., Doubeni, A.R. & Myers, A.E. Diagnosis and management of ovarian cancer. *American Family Physician* **93**, 937–944 (2016).

Eslami, S.Z., Majidzadeh, A.K., Halvaei, S., Babapirali, F. & Esmaeili, R. Microbiome and Breast Cancer: New Role for an Ancient Population. *Front Oncol* **10**, 120 (2020).

Fortner, R.T., *et al.* Sexually transmitted infections and risk of epithelial ovarian cancer: results from the

Nurses' Health Studies. *Br J Cancer* **120**, 855–860 (2019).

Gagniere, J., *et al.* Gut microbiota imbalance and colorectal cancer. *World J Gastroenterol* **22**, 501–518 (2016).

Group, N.H.W, *et al.* The NIH Human Microbiome Project. *Genome Res* **19**, 2317–2323 (2009).

Hillman, E.T., Lu, H., Yao, T. & Nakatsu, C.H. Microbial Ecology along the Gastrointestinal Tract. *Microbes Environ* **32**, 300-313 (2017).

Idahl, A., *et al.* Chlamydia trachomatis, Mycoplasma genitalium, Neisseria gonorrhoeae, human papillomavirus, and polyomavirus are not detectable in human tissue with epithelial ovarian cancer, borderline tumor, or benign conditions. *Am J Obstet Gynecol* **202**, 71 e71–76 (2010).

Idahl, A., *et al.* Serologic markers of Chlamydia trachomatis and other sexually transmitted infections and subsequent ovarian cancer risk: Results from the EPIC cohort. *Int J Cancer* (2020).

Ingerslev, K., *et al.* The prevalence of EBV and CMV DNA in epithelial ovarian cancer. *Infect Agent Cancer* **14**, 1–6 (2019).

Jacobson, D., *et al.* Shifts in gut and vaginal microbiomes are associated with cancer recurrence time in women with ovarian cancer. *PeerJ* **9**, e11574 (2021).

Jonsson, S., Oda, H., Lundin, E., Olsson, J. & Idahl, A. Chlamydia trachomatis, Chlamydial Heat Shock Protein 60 and Anti-Chlamydial Antibodies in Women with Epithelial Ovarian Tumors. *Transl Oncol* **11**, 546–551 (2018).

Kumar, L., *et al.* The Female Reproductive Tract Microbiota: Friends and Foe. *Life (Basel)* **13**(2023).

Kyo, S., Ishikawa, N., Nakamura, K. & Nakayama, K. The fallopian tube as origin of ovarian cancer: Change of diagnostic and preventive strategies. *Cancer Med* **9**, 421–431 (2020).

Labidi-Galy, S.I., *et al.* High grade serous ovarian carcinomas originate in the fallopian tube. *Nat Commun* **8**, 1093 (2017).

Laniewski, P., Ilhan, Z.E. & Herbst-Kralovetz, M.M. The microbiome and gynaecological cancer development, prevention and therapy. *Nat Rev Urol* **17**, 232–250 (2020).

Lev-Sagie, A., *et al.* The Vaginal Microbiome: II. Vaginal Dysbiotic Conditions. *J Low Genit Tract Dis* **26**, 79–84 (2022).

Li, C., *et al.* Association between vaginal microbiota and the progression of ovarian cancer. *J Med Virol* **95**, e28898 (2023).

Lisio, M.A., Fu, L., Goyeneche, A., Gao, Z.H. & Telleria, C. High-grade serous ovarian cancer: Basic sciences, clinical and therapeutic standpoints. *International Journal of Molecular Sciences* **20**(2019).

Mahoney, D.E. & Pierce, J.D. Ovarian Cancer Symptom Clusters: Use of the NIH Symptom Science Model for Precision in Symptom Recognition and Management. *Clin J Oncol Nurs* **26**, 533–542 (2022).

Mahoney, D.E., Chalise, P., Rahman, F. & Pierce, J.D. Influences of Gastrointestinal Microbiota Dysbiosis on Serum Proinflammatory Markers in Epithelial Ovarian Cancer Development and Progression. *Cancers (Basel)* **14**(2022).

Martinez, J.E., *et al.* Unhealthy Lifestyle and Gut Dysbiosis: A Better Understanding of the Effects of Poor Diet and Nicotine on the Intestinal Microbiome. *Front Endocrinol (Lausanne)* **12**, 667066 (2021).

Mathieu, K.B., Bedi, D.G., Thrower, S.L., Qayyum, A. & Bast, R.C., Jr. Screening for ovarian cancer: Imaging challenges and opportunities for improvement. *Ultrasound in Obstetrics & Gynecology* **51**, 293–303 (2018).

Medeiros, F., *et al.* The tubal fimbria is a preferred site for early adenocarcinoma in women with familial ovarian cancer syndrome. *Am J Surg Pathol* **30**, 230–236 (2006).

Miao, R., *et al.* Assessment of peritoneal microbial features and tumor marker levels as potential diagnostic tools for ovarian cancer. *PLoS One* **15**, e0227707 (2020).

Miles, S.M., Hardy, B.L. & Merrell, D.S. Investigation of the microbiota of the reproductive tract in women undergoing a total hysterectomy and bilateral salpingo-oopherectomy. *Fertil Steril* **107**, 813-820 e811 (2017).

Morikawa, A., *et al.* Altered cervicovaginal microbiota in premenopausal ovarian cancer patients. *Gene* **811**, 146083 (2022).

Muzny, C.A., *et al.* An Updated Conceptual Model on the Pathogenesis of Bacterial Vaginosis. *J Infect Dis* **220**, 1399–1405 (2019).

Nene, N.R., *et al.* Association between the cervicovaginal microbiome, BRCA1 mutation status, and risk of ovarian cancer: a case-control study. *Lancet Oncol* **20**, 1171–1182 (2019).

Neuman, H., Debelius, J.W., Knight, R. & Koren, O. Microbial endocrinology: the interplay between the microbiota and the endocrine system. *FEMS Microbiol Rev* **39**, 509–521 (2015).

Obrenovich, M.E.M. Leaky Gut, Leaky Brain? *Microorganisms* **6**(2018).

Ogunrinola, G.A., Oyewale, J.O., Oshamika, O.O. & Olasehinde, G.I. The Human Microbiome and Its Impacts on Health. *Int J Microbiol* **2020**, 8045646 (2020).

Pathak, S., Wilczynski, J.R. & Paradowska, E. Factors in Oncogenesis: Viral Infections in Ovarian Cancer. *Cancers (Basel)* **12**(2020).

Perales-Puchalt, A., *et al.* Frontline Science: Microbiota reconstitution restores intestinal integrity after cisplatin therapy. *J Leukoc Biol* **103**, 799–805 (2018).

Petersen, C. & Round, J.L. Defining dysbiosis and its influence on host immunity and disease. *Cell Microbiol* **16**, 1024–1033 (2014).

Piao, J., Lee, E.J. & Lee, M. Association between pelvic inflammatory disease and risk of ovarian cancer: An updated meta-analysis. *Gynecol Oncol* **157**, 542–548 (2020).

Qin, X., *et al.* Metagenomic analysis of the microbiome of the upper reproductive tract: combating ovarian cancer through predictive, preventive, and personalized medicine. *EPMA J* **13**, 487–498 (2022).

Ramirez-Labrada, A.G., *et al.* The Influence of Lung Microbiota on Lung Carcinogenesis, Immunity, and Immunotherapy. *Trends Cancer* **6**, 86–97 (2020).

Redondo-Useros, N., *et al.* Microbiota and Lifestyle: A Special Focus on Diet. *Nutrients* **12**(2020).

Rodrigues Hoffmann, A., Proctor, L.M., Surette, M.G. & Suchodolski, J.S. The Microbiome: The Trillions of Microorganisms That Maintain Health and Cause Disease in Humans and Companion Animals. *Vet Pathol* **53**, 10–21 (2016).

Rutkowski, M.R., *et al.* Microbially driven TLR5-dependent signaling governs distal malignant progression through tumor-promoting inflammation. *Cancer Cell* **27**, 27–40 (2015).

Schmidt, B.L., *et al.* Changes in abundance of oral microbiota associated with oral cancer. *PLoS One* **9**, e98741 (2014).

Shanmughapriya, S., *et al.* Viral and bacterial aetiologies of epithelial ovarian cancer. *Eur J Clin Microbiol Infect Dis* **31**, 2311–2317 (2012).

Siegel, R.L., Miller, K.D., Wagle, N.S. & Jemal, A. Cancer statistics, 2023. *CA Cancer J Clin* **73**, 17–48 (2023).

Sipos, A., *et al.* The role of the microbiome in ovarian cancer: mechanistic insights into oncobiosis and to bacterial metabolite signaling. *Mol Med* **27**, 33 (2021).

Smith, S.B. & Ravel, J. The vaginal microbiota, host defence and reproductive physiology. *J Physiol* **595**, 451–463 (2017).

Society, A.C. American Cancer Society. Cancer Facts & Figures 2021. (American Cancer Society, 2021).

Soong, T.R., *et al.* Evidence for lineage continuity between early serous proliferations (ESPs) in the Fallopian tube and disseminated high-grade serous carcinomas. *J Pathol* **246**, 344–351 (2018).

Sung, H., *et al.* Global Cancer Statistics 2020: GLOBOCAN Estimates of Incidence and Mortality Worldwide for 36 Cancers in 185 Countries. *CA Cancer J Clin* **71**, 209–249 (2021).

Surveillance, E., and End Results Program. SEER*Stat Database, 1975-2017. (ed. National Cancer Institute Surveillance, E., and End Results Program) (2020).

Svahn, M.F., Faber, M.T., Christensen, J., Norrild, B. & Kjaer, S.K. Prevalence of human papillomavirus in epithelial ovarian cancer tissue. A meta-analysis of observational studies. *Acta Obstet Gynecol Scand* **93**, 6–19 (2014).

Tong, J., *et al.* Changes of Intestinal Microbiota in Ovarian Cancer Patients Treated with Surgery and Chemotherapy. *Cancer Manag Res* **12**, 8125–8135 (2020).

Ursell, L.K., Metcalf, J.L., Parfrey, L.W. & Knight, R. Defining the human microbiome. *Nutr Rev* **70 Suppl 1**, S38–44 (2012).

Valdes, A.M., Walter, J., Segal, E. & Spector, T.D. Role of the gut microbiota in nutrition and health. *BMJ* **361**, k2179 (2018).

Wang, Q., *et al.* The differential distribution of bacteria between cancerous and noncancerous ovarian tissues in situ. *J Ovarian Res* **13**, 8 (2020).

Xie, X., Yang, M., Ding, Y. & Chen, J. Microbial infection, inflammation and epithelial ovarian cancer. *Oncol Lett* **14**, 1911–1919 (2017).

Xu, S., Liu, Z., Lv, M., Chen, Y. & Liu, Y. Intestinal dysbiosis promotes epithelial-mesenchymal transition by activating tumor-associated macrophages in ovarian cancer. *Pathog Dis* **77**(2019).

Xu, J., Peng, J.J., Yang, W., Fu, K. & Zhang, Y. Vaginal microbiomes and ovarian cancer: a review. *Am J Cancer Res* **10**, 743–756 (2020).

Yousefi, M., *et al.* Current insights into the metastasis of epithelial ovarian cancer - hopes and hurdles. *Cell Oncol (Dordr)* **43**, 515–538 (2020).

Zhang, M., Cheng, S., Jin, Y., Zhao, Y. & Wang, Y. Roles of CA125 in diagnosis, prediction, and oncogenesis of ovarian cancer. *Biochim Biophys Acta Rev Cancer* **1875**, 188503 (2021).

Zhao, M.F., *et al.* Novel Bacillus strains from the human gut exert anticancer effects on a broad range of malignancy types. *Invest New Drugs* **38**, 1373–1382 (2020).

Zheng, D., Liwinski, T. & Elinav, E. Interaction between microbiota and immunity in health and disease. *Cell Res* **30**, 492–506 (2020).

Zhou, B., *et al.* The biodiversity Composition of Microbiome in Ovarian Carcinoma Patients. *Sci Rep* **9**, 1691 (2019).

# 6 Insights into the Microbial Composition of Intratumoral, Reproductive Tract, and Gut Microbiota in Ovarian Cancer Patients

Qian Zhou and Qingren Meng

**Abstract**

According to the latest global cancer data, ovarian cancer is the deadliest among all gynecological malignant tumors and ranks fifth in terms of mortality. Its etiology and pathogenesis are unknown, and the 5-year survival rate of patients with advanced ovarian cancer is only 40% (Sung et al. CA Cancer J Clin 71:209–49, 2021). Recent research has shown that the human microbiota plays a crucial role in the development and progression of tumors, including ovarian cancer. Numerous studies have highlighted the complex connections between the reproductive tract microbiota, intestinal microbiota, and ovarian cancer (Jacobson et al. PeerJ 9:e11574, 2021). Therefore, this chapter will delve into composition, function, and the correlation between microbiota and immunity in the field of ovarian cancer microbiota, as well as the potential of bacteria in therapeutics and diagnostics of ovarian cancer.

Q. Zhou (✉)
International Cancer Center, Shenzhen University Medical School, Shenzhen, China
e-mail: qian.zhou@szu.edu.cn

Q. Meng
National Clinical Research Center for Infectious Diseases, Shenzhen Third People's Hospital, The Second Hospital Affiliated to Southern University of Science and Technology, Shenzhen, China

**Keywords**

Ovarian cancer · Tumor microbiota · Reproductive tract microbiota · Gut microbiota · Immune regulation

## 6.1 Introduction

In recent years, there has been a noticeable rise in the incidence of gynecological malignancies, especially among younger individuals. Although the internal factors contributing to the development and progression of these tumors are still under debate, it is widely accepted that external environmental factors, along with host gene mutations and genetics, play critical roles. Emerging evidence suggests that the reproductive tract microbiota or intestinal microbiota can influence ovarian cancer. For instance, pelvic inflammatory disease has been proven to enhance the risk of ovarian cancer, especially in severe cases. Chronic inflammation is believed to directly affect the transformation of ovarian surface epithelium into malignant epithelium, promoting the onset and progression of ovarian cancer (Lo et al. 2022; Nene et al. 2019; Lin et al. 2011). Nevertheless, the origins, exact role, and underlying mechanisms of reproductive tract microbiota, gut microbiota, and ovary tumor tissue microbiota in the emergence, development, and treatment of epithelial ovarian cancer are still



H. Schatten (ed.), *Cell and Molecular Biology of Ovarian Cancer*, Advances in Experimental Medicine and Biology 1452, https://doi.org/10.1007/978-3-031-58311-7_6

poorly understood. By examining the alterations and implications of reproductive tract microbiota and intestinal microbiota in ovarian cancer patients, we may be able to identify new avenues for comprehending the pathogenesis of this disease and designing more effective prevention strategies.

## 6.2 Overview of Ovarian Cancer Associated Microbiota

Although the reproductive tract, comprising the endometrium, fallopian tube, and pelvic cavity, has traditionally been considered a sterile environment, recent research has uncovered the presence of microbes within the reproductive tract of healthy women (Mackie et al. 1999). Interestingly, the distribution of bacterial species within this region exhibits significant variability, and changes in bacterial groups occur continuously (Backhed et al. 2015; Matamoros et al. 2013; Garcia-Grau et al. 2019). However, investigations into the interplay among reproductive tract microbiota, gut microbiota, intratumor microbiota, and ovarian cancer are still in their nascent stages. Nonetheless, these studies are groundbreaking as the utilization of 16S rDNA sequencing techniques has emerged as a potent tool in exploring human microecology, resulting in an abundance of novel discoveries. It is imperative to elucidate and authenticate the precise role and mechanisms of microbiota in the onset and progression of epithelial ovarian cancer.

### 6.2.1 Ovarian Cancer Tissue Microbial Composition and Origin

In ovarian cancer tissues, bacterial lipopolysaccharides (LPS) and lipoteichoic acid (LTA) have been detected in both cancerous and noncancerous ovarian tissues, indicating the presence of bacteria. Compared to adjacent tissues, *Bacteroides* genus has increased, whereas *Acinetobacter* genus and *Lactococcus* genus are lacking, and the abundance of *Proteobacteria* has increased. In addition, the ratio of *Proteobacteria* to *Firmicutes* has also increased in cancerous tissues, which could be beneficial for the diagnosis of ovarian cancer (Wang et al. 2020). Six highly expressed oncovirus families, including Retroviridae, Hepeviridae, HPV, Flaviviridae, Polyomaviridae, and Herpesviridae, have been found in ovarian cancer tissues.

The origin of intratumoral bacteria is currently unknown. In 2020, Walker et al. reviewed the favorable conditions for bacterial colonization within tumors, including disturbed vasculature, a hypoxic but nutrient-rich microenvironment, and immunosuppression (Xie et al. 2022; Walker et al. 2020). It is generally accepted that mucosal sites and normal adjacent tissues are important sources of the bacterial community in tumors, and hematogenous dissemination can also lead to microbial invasion (Xie et al. 2022). In the case of ovarian cancer tissues, sources of tumor microbiota could include the reproductive tract, peripheral blood, and gut.

### 6.2.2 Reproductive Tract Microbial Composition

In the endometrium, the predominant bacterial genus is *Lactobacillus*, comprising over 90% of the community (Chen et al. 2017). The bacterial community in the endometrium is similar to that of the vagina due to the proximity of the vagina to the cervix (Norenhag et al. 2020). Compared to the cervix,vagina shows a reduced *Lactobacillus* community *Lactobacillus crispatus* (CST I), *Lactobacillus gasseri* (CST II), *Lactobacillus iners* (CST III), *Lactobacillus jensenii* (CST V), and an increased anaerobic bacteria *Prevotella*, *Sneathia*, *Megasphaera*, or *Streptococcus* (CST IV) (Ravel et al. 2011; Mandal et al. 2022). Despite the diversity in species composition, the core functions remain conserved producing lactic acid, demonstrating vaginal microbial community is in a dynamic equilibrium thereby providing resilience (Ravel et al. 2011).

### 6.2.3 The Relationship Between Reproductive Tract Microbiota and Ovarian Cancer

Microbes can be found in the upper reproductive tract of healthy individuals, including the uterus, fallopian tube, and ovary. Healthy microbes in the fallopian tube and ovary have been found to suppress the development of ovarian cancer (Chen et al. 2017; Dhingra et al. 2022). However, researchers have noted that the microbial composition and proportion can be specifically altered in patients with ovarian cancer. And the α diversity of microorganisms in the upper genital tract, tumor tissue, and peritoneum has been found to decrease. There is also a difference in the abundance of certain microbial species between the upper and lower genital tracts. The upper genital tract exhibits a decrease in the abundance of *Lactobacillus*, while the proportion of *Acinetobacter* and *Pseudomonas* increases. Pathogenic microorganisms such as *Gardnerella*, *Prevotella*, *Atopobium*, *Mycoplasma*, and *Chlamydia* have been identified in ovarian cancer patients. Notably, *Brucella*, *Chlamydia*, and *Mycoplasma* have been detected in over 60% of patients with epithelial ovarian cancer, while human papillomavirus (HPV) was found in only 10% of samples. These microbial infections are considered risk factors for the development of ovarian cancer, as they contribute to the growth of atypical epithelial cells in the ovary (Idahl et al. 2020; Trabert et al. 2019a; Chen et al. 2021).

When pathogenic bacteria enter the uterus, fallopian tubes, and ovaries through the cervical or lymphatic routes, they cause microbial dysbiosis in the upper reproductive tract, leading to pelvic inflammatory disease. Women with pelvic inflammatory disease have an increased risk of developing epithelial ovarian cancer (Lo et al. 2022; Lin et al. 2011). *Chlamydia*, being the most common pathogenic microorganism in pelvic inflammatory disease, activates the paracrine *Wnt* pathway, blocks the release of mitochondrial caspase-3 and cytochrome C, escapes the attack of $CD8^+$ T cells, and inhibits apoptosis, destroying the fallopian tube epithelial cell structure and function. A positive chlamydia plasmin antibody 3 (*Pgp3*) signal is associated with an increased risk of ovarian cancer (Idahl et al. 2020; Trabert et al. 2019a). *Lactobacillus vaginalis* is a protected species of ovarian cancer. The prevalence of such species and their role in regulating vaginal pH and microbiota implies that they may act as drivers in these communities, similar to Walker's driver-passenger model. This model suggests that certain "driver" species or functional groups possess important ecological functions and considerably shape the ecosystem, while "passenger" species lack ecological impact (Walker 1992; Peterson et al. 1998). Understanding the effects of these factors on the ecology of the reproductive tract community is of significant importance in comprehending the stability, resistance, and resilience of the reproductive tract community. This will eventually assist in formulating strategies to maintain the health of the human reproductive tract and prevent reproductive tract tumors, particularly ovarian cancer.

### 6.2.4 Gut Microbiota in Ovarian Cancer Patients and Animal Models

The normal gut microbiome is a dynamic and complex system that can be influenced by host characteristics such as race, sex, age, hormone levels, and immune response, as well as environmental factors including diet, smoking, and medications. A healthy gut microbiota is comprised of five bacterial phyla, including *Firmicutes*, *Bacteroidetes*, *Actinobacteria*, *Proteobacteria*, and *Fusobacteria*, which together make up roughly 90% of the total gut microbiota (Erdman and Poutahidis 2017; Bodai and Nakata, 2020; Hamada et al. 2019; Nejman et al. 2020; Al-Hilu and Al-Shujairi 2020; Fassarella et al. 2021; Soto-Pantoja et al. 2021).

Abdominal discomfort is a characteristic symptom of ovarian cancer, while gastrointestinal adverse reactions are common during ovarian cancer clinical treatment. Therefore, it is crucial to investigate the relationship between intestinal microbiota and ovarian cancer. Dysregulation of gut microbes has been shown to be closely associated with the development of ovarian cancer, promotion of tumor growth, and occurrence of epithelial-mesenchymal transition in xenograft mice (Laniewski et al. 2020). Furthermore, depletion of gut microbiota by antibiotics significantly accelerates the growth rate of ovarian cancer in mice bearing human ovarian adenocarcinoma SKOV-4 cells, indicating that an imbalance of gut microbiota promotes the progression of ovarian cancer (Xu et al. 2019).

## 6.3 Functions of Ovarian Cancer-Associated Microbiota Metabolites

Bacterial metabolites and components can have local effects or enter the host's systemic circulation and exert hormone-like effects at a distance. In this section, we will examine bacterial metabolites and their role in ovarian cancer pathogenesis, hormone regulation, and inflammatory and immune regulation.

### 6.3.1 Microbial Metabolites Affect Ovarian Cancer Development

Microbial metabolites, produced by both the healthy eubiome and the oncobiome, have demonstrated the ability to exert autocrine, paracrine, and hormone-like effects. This was evidenced in cases of many cancer types including breast cancer, pancreatic adenocarcinoma, colorectal cancer, gastric cancer, and hepatocellular carcinoma (Miko et al. 2019; Kiss et al. 2020; Kuo et al. 2016; Shellman et al. 2017; Ravnik et al. 2021; Sittipo et al. 2019). Lipopolysaccharides, lysolecithin, tryptophan, and their derivatives have clear effects on ovarian cancer, while the roles of other metabolites such as short-chain fatty acids, biogenic amines, and secondary bile acids in ovarian cancer are not yet fully understood (Sipos et al. 2021). The proportion of gram-negative bacteria in the tumor tissues of ovarian cancer patients increases, and endogenous pro-inflammatory components are upregulated. LPS, the component of gram-negative bacteria, is a pathogen-associated molecular pattern (PAMP); can stimulate TLR2, TLR4, and TLR5; activate the PI3K signaling pathway, matrix metalloproteinase-related family MMP2 and MMP9 expression; activate cancer cells and tumor-associated macrophages (TAM); and induce epithelial-mesenchymal transition (Park et al. 2017; Szajnik et al. 2009; Xu et al. 1995). Lysolecithin and lysophosphatidylserine have been shown to induce protein kinase B (Akt), mitogen-activated protein kinase (MAPK), and Ca2+ signaling pathways; upregulate the expression of vascular endothelial growth factor; and induce ovarian cancer cell proliferation, migration, and invasion (Borella et al. 2021; Estrella et al. 2007). Certain bacterial species can breach the epithelial barrier and directly introduce toxins into host cells, including colibactin, which damages DNA, modifying tumor microenvironment, and provokes carcinogenic reactions (Lucas et al. 2020; Wilson et al. 2019; Dalmasso et al. 2014).

Small nutrient molecules in the gut are recognized as crucial players in the pathogenesis of ovarian cancer, specifically a reduction in serum levels of serotonin and indole-3-acetic acid, which have been observed in ovarian cancer patients and found to be inversely associated with poor prognosis. Serotonin is an essential energy source that supports *Lactobacilli* growth while simultaneously suppressing the expansion of pathogenic bacteria (Sonner et al. 2019; Turkoglu et al. 2016). If dietary tryptophan supply is not provided, there may be an impact on trp metabolism through the 5-hydroxytryptamin (5-HT)/serotonin pathway. The metabolism of L-tryptophan was found to be altered in patients with OC, as evidenced by decreased plasma levels of tryptophan and its metabolites, 3-indolepropionic acid and 5-hydroxyindoleacetaldehyde (Turkoglu et al. 2016; Ke et al. 2015). Moreover, serotonin activates aryl-hydrocarbon receptor (AHR) and pregnane X receptor (PXR) to regulate mucosal immunity (Rossi et al. 2020).

### 6.3.2 Microbiota Affect Hormone Regulation and Carcinogenesis

The relationship between estrogen and microbiota has been confirmed by contemporary scientific research, which suggests that this linkage has a role in the initiation and progression of ovarian cancer. According to a nested case-control study, higher levels of circulating estrogens significantly increase the risk of non-mucinous ovarian cancer. In women who are not taking menopausal hormones during the blood sampling period, there is a clear correlation between elevated estrogen levels and a significantly higher susceptibility to non-serous ovarian cancer (Trabert et al. 2019b). Increased estrogen levels mainly promote gene transcription through direct receptor binding and indirectly trigger other transcription factors that regulate cellular proliferation, differentiation, and apoptosis (Laniewski et al. 2020; Alpuim Costa et al. 2021). The diversity of intestinal microbiota in postmenopausal women increased with the proportion of urinary estrogen metabolites (Fuhrman et al. 2014). Gut microbiota's Firmicutes and Bacteroides secrete β-glucuronidase and β-glucosidase, respectively, which participate in the estrogen-gut microbiome axis (Baker et al. 2017) and enhance its potential to induce carcinogenesis (Parida and Sharma 2019; Yang et al. 2020).

### 6.3.3 Microbiota Regulate Inflammatory and Immune Regulation

An increased presence of microorganisms, an increase in pro-inflammatory factors, and a decrease in immune suppressive pathways in ovarian cancer tissues suggest a relationship between microbiota-triggered inflammation and immune regulation. Toll-like receptors (TLR) enable dendritic cells, which are essential to the immune response, to detect microbial presence and produce a variety of cytokines that regulate T-cell differentiation, creating a more balanced response to bacterial antigens (Matson et al. 2021; Jang et al. 2011; Tanoue et al. 2019). However, if the microbial dysbiosis persists, immune cells are activated and inflammatory factors are released, leading to a heightened inflammatory response. This response can contribute to an accumulation of DNA damage and mutations, ultimately promoting the development of ovarian cancer.

TLR4 can induce an inflammatory response through MyD88 and MAL upon connecting with LPS and subsequently activate NF-κB, IL-6, and VEGF. This can lead to enhanced proliferation of cancer cells and increased resistance to antitumor medicines (Szajnik et al. 2009). IL-6's oncogenic effects are particularly relevant to ovarian cancer, as it triggers positive feedback loops that induce JAK and Stat3 transcription (Browning et al. 2018). TLR5, on the other hand, is closely associated with immune suppression of microbiota, along with tumor inflammation and metastasis. Its interaction with bacterial flagellin induces activation of the MyD88/TRAF6/NF-κB pathway, producing more cytokines like interferon α/β and IL-6, which lead to a quicker increase in tumor size (Park and Kim 2017). This cancer-promoting effect depends on the gut microbiota, and its growth-promoting effect is lost when the gut bacteria are eliminated (Rutkowski et al. 2015).

## 6.4 Bacteria in Therapeutics and Diagnostics of Ovarian Cancer

Microbes play a dual role in cancer clinical therapy. On the one hand, they and their metabolites can enhance the immune system and improve antitumor activity. On the other hand, some microbial metabolites can weaken the therapeutic effectiveness of drugs. Furthermore, alterations in the microbiome can contribute to the initiation and advancement of cancer. Similarly, some cancers can disrupt and change microbial composition. Bacterial dysbiosis and metabolite signaling are detectable in blood, urine, feces, and cervicovaginal discharges/rinses, and can be used for both diagnostics and therapeutics of ovarian cancer. Studies suggest that microbiome signatures in intratumoral regions, genital tracts, serum, or peritoneum can be utilized as biomarkers for diagnosing ovarian cancer (Wang et al. 2020; Poore et al. 2020; Zhou et al. 2019; Banerjee et al. 2017; Kim et al. 2020; Mori et al. 2019). Moreover, metabolomics studies have also revealed available biomarkers (Turkoglu et al. 2016). These biomarkers can aid in early diagnosis, screening, risk assessment, and prognosis. Since we now have a better understanding of bacterial metabolite signaling, further research is necessary to explore its practical use.

### 6.4.1 The Role of Microbiota in Therapeutic Effects in Ovarian Cancer

Chemotherapy and immunotherapy are commonly used in treating ovarian cancer, but research shows that chemotherapy resistance leads to relapse in up to 80% of advanced-stage ovarian cancer patients within 2 years (Dood et al. 2018; Coleman et al. 2019). The variability in patients' response to immunotherapy presents a significant hurdle (Hanggi and Ruffell 2023). Recent studies highlight how microbiota can have both positive and negative effects on tumor chemotherapy and targeted therapy (Serna et al. 2020; Chambers et al. 2022). On the one hand, microbiota can trigger chemotherapy resistance, while on the other hand, it can help enhance antitumor activity. Nevertheless, the effect of microbiota in clinical cancer treatment is still debated. We will discuss the double-edged role of microbiota in cancer treatment.

#### 6.4.1.1 The Potential Role of Microbiota in Ovarian Cancer Therapy Drug Resistance

The colonization of gram-negative bacteria in ovarian cancer tissues activates the pattern recognition receptors on the surface of innate immune cells and downstream signaling cascades, leading to an increase in metastasis and drug resistance of ovarian cancer. Moreover, the C-type lectin receptor (CLR) induces tumor-associated macrophages in ovarian cancer tissues to participate in immune escape (Allavena et al. 2010). Furthermore, the downregulation of multidrug resistance proteins involved in paclitaxel resistance is observed following TLR4 inactivation, further highlighting the potential influence of microbiota on chemotherapy resistance in ovarian cancer (Zandi et al. 2019).

Nevertheless, certain bacterial species contribute to reversed resistance. Fecal microbiota transplantation and probiotic supplementation have been widely used in the treatment of nonmalignant diseases. The treatment of platinum-based and anti-PD-1 monoclonal antibody, combined with probiotics, can significantly enhance the efficacy of ovarian cancer. Furthermore, the supplementation with *Akkermansia* or implantation of fecal bacteria from drug-responsive donors can reverse resistance to anti-PD-1 therapy (Routy et al. 2018). *Akkermansia* supplementation can also reverse the accelerated tumor growth of ovarian-cancer-bearing mice, accompanied by the accumulation of acetate and increased interferon $\gamma$ (IFN$\gamma$) secretion of CD8+ T cells and its tumor-killing property (Wang et al. 2022). These studies emphasize the importance of microbiota in ovarian cancer immune surveillance during cancer development and immunotherapy.

#### 6.4.1.2 A Dual Role of Antibiotic Therapy for Ovarian Cancer

Antibiotic therapy for ovarian cancer has shown potential, although some studies have yielded inconsistent results. Modulating the structure of microbiota presents an opportunity to mitigate the adverse reactions of chemotherapy and enhance treatment efficacy. For example, studies in mice have shown that reconstitution of intestinal microbes after cisplatin treatment can expedite intestinal healing and that *E. coli* Nissle1917 can alleviate mucositis induced by topoisomerase inhibitors. Additionally, anticancer drugs may stimulate gut microbes to migrate to secondary lymphatic organs, where they induce an immune response. For example, the combination of chloramphenicol, salinomycin, and cisplatin inhibits tumor growth. Anthracyclines and cyclophosphamide can promote the loss of intestinal barrier and bacterial translocation (Yang et al. 2013), thereby stimulating helper and memory T cells (Ahlmann and Hempel 2016), suggesting that the gut microbiome plays a favorable role in promoting antitumor effects and immune response to chemotherapy.

Nevertheless, antibiotic therapy may have adverse effects on the progression of ovarian cancer. For instance, a study of 402 patients with stage III/IV epithelial ovarian cancer found that patients treated with antibiotics for more than 48 h during chemotherapy had lower progression-free survival and overall survival compared to patients not treated with antibiotics, particularly against gram-positive bacteria. Furthermore, the combination of ampicillin, vancomycin, neomycin, and metronidazole promoted the growth and invasion of transplanted tumors in nude mice (Iida et al. 2013). Cheng et al. reviewed that antibiotic therapy can affect intestinal microbiota, thus regulating the polarization of the immune system and promoting ovarian cancer progression through inflammation (Cheng et al. 2020). Research has shown that transplanting fecal microbiota from non-small-cell lung cancer patients into tumor-bearing mice after antibiotic intervention enhances the anti-tumor effect of PD-1 blockade (Routy et al. 2018). Additionally, in a recent study, researchers traced the gut bacteria and metabolites in rectal cancer patients during neoadjuvant chemoradiotherapy and identified *Bacteroides vulgatus* as a driver bacterium of therapeutic resistance by supplying tumors with nucleotides (Teng et al. 2023; Wong et al. 2023). Additional studies are required to evaluate the advantages and disadvantages of using antibiotics to treat ovarian cancer. Examining how antibiotic therapy affects the microbiota of ovarian cancer could provide a better understanding of its exact role in tumor formation and assist in the advancement of improved prevention and treatment strategies.

Additionally, activating the immune system through microbial composition and immunogenicity could be successful as a therapeutic approach. New protocols for adoptive cellular immunotherapy, including CAR-T and dendritic cell vaccines, are designed to enhance immune checkpoint blocking (Caro et al. 2022; Zhang et al. 2020; Guo et al. 2020). Vaginal microbiome transplantation has also shown potential in treating bacterial vaginosis, leading to significant relief of symptoms (Lev-Sagie et al. 2019). To optimize the delivery of probiotics into the vagina, various novel methods have been proposed, including incorporating them into fibers to improve their survival (Zupancic et al. 2019) or using 3D-printed scaffolds embedded with bacteria (Chandrashekhar et al. 2021). Two recent reviews have explored vaginal microbiota transplantation and its potential application in treating vaginal bacterial infections (Han et al. 2021; Junca et al. 2022). However, the role of vaginal microbiota transplantation in vaginitis-related ovarian cancer requires further investigation. While personalized immunotherapy for ovarian cancer still faces challenges, these strategies demonstrate promising potential.

### 6.4.2 Nutrient and Diet Modify Microbiome and Help Prevent Ovarian Cancer

The correlation between diet and cancer has been identified. Prior to the widespread use of antibiotics, diet played a critical role in the prevention of infectious diseases, which indicates its potential role in preventing cancer. Diet and nutrients have the ability to regulate the composition of the female reproductive tract and intestinal microbiota, influence gene expression through genetic and epigenetic inheritance, and affect disease susceptibility. Studies have shown that the ketogenic, Mediterranean, and vegetarian diets are all beneficial in regard to the prognosis of ovarian cancer (Cohen et al. 2018; Wallace et al. 2019). The ketogenic diet involves high fat consumption and low carbohydrate and protein intake, prompting a switch in metabolism from glucose to ketones, which can decrease inflammation, hinder tumor angiogenesis, and boost cell apoptosis (Wallace et al. 2019). A high intake of saturated fatty acids, animal fats, and trans fatty acids can raise the odds of ovarian cancer, while vegetables, fruits, fiber, vitamin C, and polyunsaturated fatty acids are protective factors. The Mediterranean diet focuses on legumes, nuts, fruits, whole grains, and vegetables with limited red meat intake, utilizing fat from virgin olive oil. Similarly, the vegetarian diet has high levels of fiber and phytoestrogens, which can increase microbiota diversity and produce anticancer effects (Playdon et al. 2017). Other nutrients such as naringenin have been recognized for suppressing mouse epithelial ovarian cancer by inhibiting proliferation and modulating gut microbiota, with a significant increase in the abundances of *Alistipes* and *Lactobacillus* (Lin et al. 2022).

## 6.5 Conclusion and Future Perspectives

In recent years, high-throughput sequencing, quantitative PCR, and live bacteria culturing technologies have led to the discovery of distinct bacterial communities in the female reproductive tract. These communities form a continuous microbiota that changes from the vagina to the ovaries. Scientists have not only discovered a correlation between ovarian cancer and the microbiome, but they have also made substantial strides in identifying and understanding the microorganisms present in the female reproductive tract. The dysbiosis of female reproductive tract may cause DNA damage, activate cancer-promoting signaling pathways, and change the immune microenvironment, thereby promoting the progression of ovarian cancer (Fig. 6.1). The unique microbiota and metabolites found in tumor tissues, the reproductive tract, the intestinal tract, and the peritoneum of ovarian cancer patients can serve as biomarkers for early diagnosis and offer new strategies for screening, treatment stratification, and prognosis.

Research has shown that the microbiota of the female reproductive system plays a crucial role in maintaining the host's health by protecting against pathogenic bacteria, regulating the immune system, and influencing hormonal processes such as estrogen production. Disruptions to the female microbiome may contribute to the development of ovarian cancer. Specifically, imbalances in the microbiota could lead to chronic inflammation, DNA damage, and

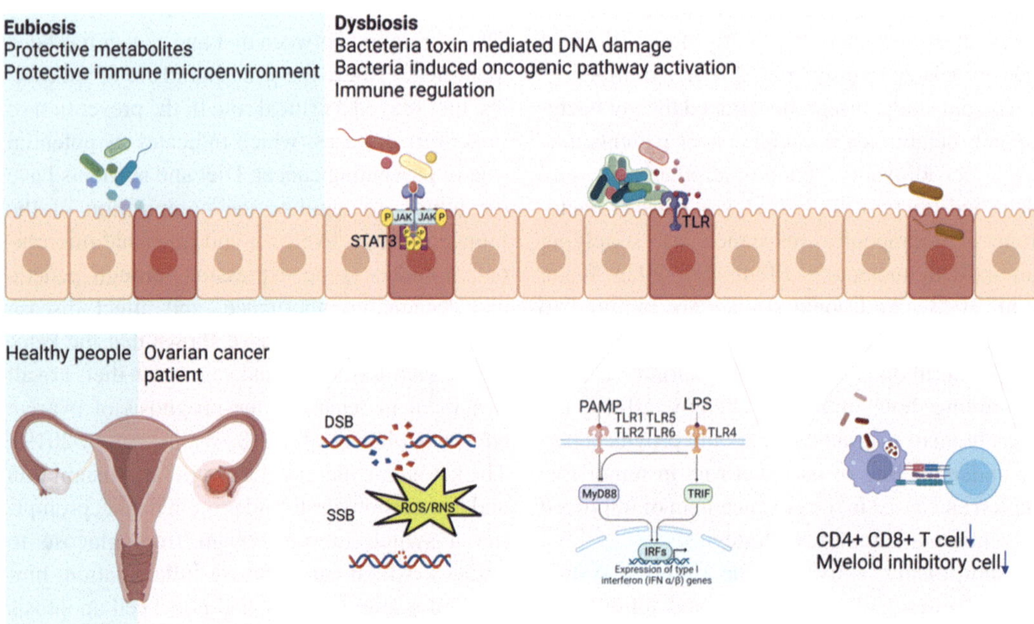


**Fig. 6.1** Simplified schematic of the various processes involved in the progression of bacteria-induced tumorigenesis

impaired immune function. However, the specific microorganisms involved and the causal relationship between the microbiota and ovarian cancer are not yet fully understood, and the clinical value and feasibility of microbial-based approaches require further investigation. Therefore, comprehensive investigation of ovarian cancer pathogenesis that includes host genetics, epigenetics, microbiota, lifestyle, and other environmental factors is essential to understand the underlying mechanisms.

Scientists are exploring various methods to manipulate the microbiome as a potential avenue for ovarian cancer prevention and treatment, including probiotics, prebiotics, and fecal microbiota transplant therapies. Demonstrating the presence of microbiota in traditionally uncharted territories remains a challenging task. Combining high-throughput sequencing, quantitative PCR, live bacteria culturing, and other techniques may make a stronger case, as illustrated here. It is also intriguing to consider the possibility of endosymbiotic bacteria, rather than free-living bacteria, being involved in the process of ovarian cancer carcinogenesis, and in turn the invasion of bacteria during ovarian carcinogenesis. However, further investigation is needed to confirm this possibility.

## References


Sung H, Ferlay J, Siegel R L, et al. Global Cancer Statistics 2020: GLOBOCAN Estimates of Incidence and Mortality Worldwide for 36 Cancers in 185 Countries [J]. CA Cancer J Clin, 2021, 71(3): 209–49.

Jacobson D, Moore K, Gunderson C, et al. Shifts in gut and vaginal microbiomes are associated with cancer recurrence time in women with ovarian cancer [J]. PeerJ, 2021, 9: e11574.

Lo H W, Weng S F, Chen H S, et al. Pelvic inflammatory disease is associated with ovarian cancer development in women with endometriosis: A cohort study in Taiwan [J]. Int J Gynecol Obstet, 2022, 158(1): 145–52.

Nene N R, Reisel D, Leimbach A, et al. Association between the cervicovaginal microbiome, BRCA1 mutation status, and risk of ovarian cancer: a case-control study [J]. Lancet Oncol, 2019, 20(8): 1171–82.

Lin H W, Tu Y Y, Lin S Y, et al. Risk of ovarian cancer in women with pelvic inflammatory disease: a population-based study [J]. Lancet Oncol, 2011, 12(9): 900–4.

Mackie R I, Sghir A, Gaskins H R. Developmental microbial ecology of the neonatal gastrointestinal tract [J]. Am J Clin Nutr, 1999, 69(5): 1035S–45S.

Backhed F, Roswall J, Peng Y Q, et al. Dynamics and Stabilization of the Human Gut Microbiome during the First Year of Life [J]. Cell Host Microbe, 2015, 17(5): 690–703.

Matamoros S, Gras-Leguen C, Le Vacon F, et al. Development of intestinal microbiota in infants and its impact on health [J]. Trends Microbiol, 2013, 21(4): 167–73.

Garcia-Grau I, Simon C, Moreno I. Uterine microbiome-low biomass and high expectationsdagger [J]. Biol Reprod, 2019, 101(6): 1102–14.

Wang Q, Zhao L B, Han L, et al. The differential distribution of bacteria between cancerous and noncancerous ovarian tissues in situ [J]. J Ovarian Res, 2020, 13(1).

Xie Y, Xie F, Zhou X, et al. Microbiota in Tumors: From Understanding to Application [J]. Adv Sci (Weinh), 2022, 9(21): e2200470.

Walker S P, Tangney M, Claesson M J. Sequence-Based Characterization of Intratumoral Bacteria-A Guide to Best Practice [J]. Front Oncol, 2020, 10: 179.

Chen C, Song X, Wei W, et al. The microbiota continuum along the female reproductive tract and its relation to uterine-related diseases [J]. Nat Commun, 2017, 8(1): 875.

Norenhag J, Du J, Olovsson M, et al. The vaginal microbiota, human papillomavirus and cervical dysplasia: a systematic review and network meta-analysis [J]. BJOG, 2020, 127(2): 171–80.

Ravel J, Gajer P, Abdo Z, et al. Vaginal microbiome of reproductive-age women [J]. Proc Natl Acad Sci U S A, 2011, 108 Suppl 1(Suppl 1): 4680–7.

Mandal S, Bandyopadhyay S, Tyagi K, et al. Human microbial dysbiosis as driver of gynecological malignancies [J]. Biochimie, 2022, 197: 86–95.

Dhingra A, Sharma D, Kumar A, et al. Microbiome and Development of Ovarian Cancer [J]. Endocr, Metab Immune Disord: Drug Targets, 2022, 22(11): 1073–90.

Idahl A, Le Cornet C, Gonzalez Maldonado S, et al. Serologic markers of Chlamydia trachomatis and other sexually transmitted infections and subsequent ovarian cancer risk: Results from the EPIC cohort [J]. Int J Cancer, 2020, 147(8): 2042–52.

Trabert B, Waterboer T, Idahl A, et al. Antibodies Against Chlamydia trachomatis and Ovarian Cancer Risk in Two Independent Populations [J]. J Natl Cancer Inst, 2019a, 111(2): 129–36.

Chen S, Li Y, Qian L, et al. A Review of the Clinical Characteristics and Novel Molecular Subtypes of Endometrioid Ovarian Cancer [J]. Front Oncol, 2021, 11: 668151.

Walker B H. Biodiversity and Ecological Redundancy [J]. Conserv Biol, 1992, 6(1): 18–23.

Peterson G, Allen C R, Holling C S. Ecological resilience, biodiversity, and scale [J]. Ecosystems, 1998, 1(1): 6–18.

Erdman S E, Poutahidis T. Gut microbiota modulate host immune cells in cancer development and growth [J]. Free Radic Biol Med, 2017, 105: 28–34.

Bodai B I, Nakata T E. Breast Cancer: Lifestyle, the Human Gut Microbiota/Microbiome, and Survivorship [J]. Perm J, 2020, 24.

Hamada T, Nowak J A, Milner D A, et al. Integration of microbiology, molecular pathology, and epidemiology: a new paradigm to explore the pathogenesis of microbiome-driven neoplasms [J]. J Pathol, 2019, 247(5): 615–28.

Nejman D, Livyatan I, Fuks G, et al. The human tumor microbiome is composed of tumor type-specific intracellular bacteria [J]. Science, 2020, 368(6494): 973–80.

Al-Hilu S A, Al-Shujairi W H. Dual Role of Bacteria in Carcinoma: Stimulation and Inhibition [J]. Int J Microbiol, 2020, 2020: 4639761.

Fassarella M, Blaak E E, Penders J, et al. Gut microbiome stability and resilience: elucidating the response to perturbations in order to modulate gut health [J]. Gut, 2021, 70(3): 595–605.

Soto-Pantoja D R, Gaber M, Arnone A A, et al. Diet alters entero-mammary signaling to regulate the breast microbiome and tumorigenesis [J]. Cancer Res, 2021.

Laniewski P, Ilhan Z E, Herbst-Kralovetz M M. The microbiome and gynaecological cancer development, prevention and therapy [J]. Nat Rev Urol, 2020, 17(4): 232–50.

Xu S, Liu Z, Lv M, et al. Intestinal dysbiosis promotes epithelial-mesenchymal transition by activating tumor-associated macrophages in ovarian cancer [J]. Pathog Dis, 2019, 77(2).

Miko E, Kovacs T, Sebo E, et al. Microbiome-Microbial Metabolome-Cancer Cell Interactions in Breast Cancer-Familiar, but Unexplored [J]. Cells-Basel, 2019, 8(4).

Kiss B, Miko E, Sebo E, et al. Oncobiosis and Microbial Metabolite Signaling in Pancreatic Adenocarcinoma [J]. Cancers (Basel), 2020, 12(5).

Kuo W T, Lee T C, Yu L C H. Eritoran Suppresses Colon Cancer by Altering a Functional Balance in Toll-like Receptors That Bind Lipopolysaccharide [J]. Cancer Res, 2016, 76(16): 4684–95.

Shellman Z, Aldhahrani A, Verdon B, et al. Bile acids: a potential role in the pathogenesis of pharyngeal malignancy [J]. Clin Otolaryngol, 2017, 42(5): 969–73.

Ravnik Z, Muthiah I, Dhanaraj P. Computational studies on bacterial secondary metabolites against breast cancer [J]. J Biomol Struct Dyn, 2021, 39(18): 7056–64.

Sittipo P, Shim J W, Lee Y K. Microbial Metabolites Determine Host Health and the Status of Some Diseases [J]. Int J Mol Sci, 2019, 20(21).

Sipos A, Ujlaki G, Miko E, et al. The role of the microbiome in ovarian cancer: mechanistic insights into oncobiosis and to bacterial metabolite signaling [J]. Mol Med, 2021, 27(1).

Park G B, Chung Y H, Kim D. Induction of galectin-1 by TLR-dependent PI3K activation enhances epithelial-mesenchymal transition of metastatic ovarian cancer cells [J]. Oncol Rep, 2017, 37(5): 3137–45.

Szajnik M, Szczepanski M J, Czystowska M, et al. TLR4 signaling induced by lipopolysaccharide or paclitaxel regulates tumor survival and chemoresistance in ovarian cancer [J]. Oncogene, 2009, 28(49): 4353–63.

Xu Y, Fang X J, Casey G, et al. Lysophospholipids activate ovarian and breast cancer cells [J]. Biochem J, 1995, 309 ( Pt 3)(Pt 3): 933–40.

Borella F, Carosso A R, Cosma S, et al. Gut Microbiota and Gynecological Cancers: A Summary of Pathogenetic Mechanisms and Future Directions [J]. ACS Infect Dis, 2021, 7(5): 987–1009.

Estrella V C, Eder A M, Liu S, et al. Lysophosphatidic acid induction of urokinase plasminogen activator secretion requires activation of the p38MAPK pathway [J]. Int J Oncol, 2007, 31(2): 441–9.

Lucas C, Salesse L, Hoang M H T, et al. Autophagy of Intestinal Epithelial Cells Inhibits Colorectal Carcinogenesis Induced by Colibactin-Producing Escherichia coil in Apc(Min/+) Mice [J]. Gastroenterology, 2020, 158(5): 1373–88.

Wilson M R, Jiang Y, Villalta P W, et al. The human gut bacterial genotoxin colibactin alkylates DNA [J]. Science, 2019, 363(6428).

Dalmasso G, Cougnoux A, Delmas J, et al. The bacterial genotoxin colibactin promotes colon tumor growth by modifying the tumor microenvironment [J]. Gut Microbes, 2014, 5(5): 675-80.

Sonner J K, Keil M, Falk-Paulsen M, et al. Dietary tryptophan links encephalogenicity of autoreactive T cells with gut microbial ecology [J]. Nat Commun, 2019, 10(1): 4877.

Turkoglu O, Zeb A, Graham S, et al. Metabolomics of biomarker discovery in ovarian cancer: a systematic review of the current literature [J]. Metabolomics, 2016, 12(4).

Ke C, Hou Y, Zhang H, et al. Large-scale profiling of metabolic dysregulation in ovarian cancer [J]. Int J Cancer, 2015, 136(3): 516–26.

Rossi T, Vergara D, Fanini F, et al. Microbiota-Derived Metabolites in Tumor Progression and Metastasis [J]. Int J Mol Sci, 2020, 21(16).

Trabert B, Coburn S B, Falk R T, et al. Circulating estrogens and postmenopausal ovarian and endometrial cancer risk among current hormone users in the Women's Health Initiative Observational Study [J]. Cancer Causes Control, 2019b, 30(11): 1201–11.

Alpuim Costa D, Nobre J G, Batista M V, et al. Human Microbiota and Breast Cancer-Is There Any Relevant Link?-A Literature Review and New Horizons Toward Personalised Medicine [J]. Front Microbiol, 2021, 12: 584332.

Fuhrman B J, Feigelson H S, Flores R, et al. Associations of the Fecal Microbiome With Urinary Estrogens and Estrogen Metabolites in Postmenopausal Women [J]. J Clin Endocrinol Metab, 2014, 99(12): 4632–40.

Baker J M, Al-Nakkash L, Herbst-Kralovetz M M. Estrogen-gut microbiome axis: Physiological and clinical implications [J]. Maturitas, 2017, 103: 45–53.

Parida S, Sharma D. The Microbiome-Estrogen Connection and Breast Cancer Risk [J]. Cells-Basel, 2019, 8(12).

Yang X Y, Lu D, Zhuo J Y, et al. The Gut-liver Axis in Immune Remodeling: New insight into Liver Diseases [J]. International Journal of Biological Sciences, 2020, 16(13): 2357–66.

Matson V, Chervin C S, Gajewski T F. Cancer and the Microbiome-Influence of the Commensal Microbiota on Cancer, Immune Responses, and Immunotherapy [J]. Gastroenterology, 2021, 160(2): 600–13.

Jang M J, Kim J E, Chung Y H, et al. Dendritic cells stimulated with outer membrane protein A (OmpA) of Salmonella typhimurium generate effective anti-tumor immunity [J]. Vaccine, 2011, 29(13): 2400–10.

Tanoue T, Morita S, Plichta D R, et al. A defined commensal consortium elicits CD8 T cells and anti-cancer immunity [J]. Nature, 2019, 565(7741): 600–5.

Browning L, Patel M R, Horvath E B, et al. IL-6 and ovarian cancer: inflammatory cytokines in promotion of metastasis [J]. Cancer Manag Res, 2018, 10: 6685–93.

Park G B, Kim D. TLR5/7-mediated PI3K activation triggers epithelial-mesenchymal transition of ovarian cancer cells through WAVE3-dependent mesothelin or OCT4/SOX2 expression [J]. Oncol Rep, 2017, 38(5): 3167–76.

Rutkowski M R, Stephen T L, Svoronos N, et al. Microbially driven TLR5-dependent signaling governs distal malignant progression through tumor-promoting inflammation [J]. Cancer Cell, 2015, 27(1): 27–40.

Poore G D, Kopylova E, Zhu Q, et al. Microbiome analyses of blood and tissues suggest cancer diagnostic approach [J]. Nature, 2020, 579(7800): 567–74.

Zhou B, Sun C, Huang J, et al. The biodiversity Composition of Microbiome in Ovarian Carcinoma Patients [J]. Sci Rep, 2019, 9(1): 1691.

Banerjee S, Tian T, Wei Z, et al. The ovarian cancer oncobiome [J]. Oncotarget, 2017, 8(22): 36225–45.

Kim S I, Kang N, Leem S, et al. Metagenomic Analysis of Serum Microbe-Derived Extracellular Vesicles and Diagnostic Models to Differentiate Ovarian Cancer and Benign Ovarian Tumor [J]. Cancers (Basel), 2020, 12(5).

Mori G, Orena B S, Cultrera I, et al. Gut Microbiota Analysis in Postoperative Lynch Syndrome Patients [J]. Front Microbiol, 2019, 10: 1746.

Dood R L, Zhao Y, Armbruster S D, et al. Defining Survivorship Trajectories Across Patients With Solid Tumors: An Evidence-Based Approach [J]. JAMA Oncol, 2018, 4(11): 1519–26.

Coleman R L, Spirtos N M, Enserro D, et al. Secondary Surgical Cytoreduction for Recurrent Ovarian Cancer [J]. N Engl J Med, 2019, 381(20): 1929–39.

Hanggi K, Ruffell B. Cell death, therapeutics, and the immune response in cancer [J]. Trends Cancer, 2023.

Serna G, Ruiz-Pace F, Hernando J, et al. Fusobacterium nucleatum persistence and risk of recurrence after preoperative treatment in locally advanced rectal cancer [J]. Ann Oncol, 2020, 31(10): 1366–75.

Chambers L M, Esakov Rhoades E L, Bharti R, et al. Disruption of the Gut Microbiota Confers Cisplatin Resistance in Epithelial Ovarian Cancer [J]. Cancer Res, 2022, 82(24): 4654–69.

Allavena P, Chieppa M, Bianchi G, et al. Engagement of the mannose receptor by tumoral mucins activates an immune suppressive phenotype in human tumor-associated macrophages [J]. Clin Dev Immunol, 2010, 2010: 547179.

Zandi Z, Kashani B, Poursani E M, et al. TLR4 blockade using TAK-242 suppresses ovarian and breast cancer cells invasion through the inhibition of extracellular matrix degradation and epithelial-mesenchymal transition [J]. Eur J Pharmacol, 2019, 853: 256–63.

Routy B, Le Chatelier E, Derosa L, et al. Gut microbiome influences efficacy of PD-1-based immunotherapy against epithelial tumors [J]. Science, 2018, 359(6371): 91–7.

Wang Z, Qin X, Hu D, et al. Akkermansia supplementation reverses the tumor-promoting effect of the fecal microbiota transplantation in ovarian cancer [J]. Cell Rep, 2022, 41(13): 111890.

Yang J, Liu K X, Qu J M, et al. The changes induced by cyclophosphamide in intestinal barrier and microflora in mice [J]. Eur J Pharmacol, 2013, 714(1-3): 120–4.

Ahlmann M, Hempel G. The effect of cyclophosphamide on the immune system: implications for clinical cancer therapy [J]. Cancer Chemother Pharmacol, 2016, 78(4): 661–71.

Iida N, Dzutsev A, Stewart C A, et al. Commensal bacteria control cancer response to therapy by modulating the tumor microenvironment [J]. Science, 2013, 342(6161): 967–70.

Cheng H, Wang Z, Cui L, et al. Opportunities and Challenges of the Human Microbiome in Ovarian Cancer [J]. Front Oncol, 2020, 10: 163.

Teng H, Wang Y, Sui X, et al. Gut microbiota-mediated nucleotide synthesis attenuates the response to neoadjuvant chemoradiotherapy in rectal cancer [J]. Cancer Cell, 2023, 41(1): 124–38 e6.

Wong C C, Fong W, Yu J. Gut microbes promote chemoradiotherapy resistance via metabolic cross-feeding [J]. Cancer Cell, 2023, 41(1): 12–4.

Caro A A, Deschoemaeker S, Allonsius L, et al. Dendritic Cell Vaccines: A Promising Approach in the Fight against Ovarian Cancer [J]. Cancers (Basel), 2022, 14(16).

Zhang X, He T, Li Y, et al. Dendritic Cell Vaccines in Ovarian Cancer [J]. Front Immunol, 2020, 11: 613773.

Guo Q, Yang Q, Li J, et al. Advanced clinical trials of dendritic cell vaccines in ovarian cancer [J]. J Investig Med, 2020, 68(7): 1223–7.

Lev-Sagie A, Goldman-Wohl D, Cohen Y, et al. Vaginal microbiome transplantation in women with intractable bacterial vaginosis [J]. Nat Med, 2019, 25(10): 1500–4.

Zupancic S, Skrlec K, Kocbek P, et al. Effects of Electrospinning on the Viability of Ten Species of Lactic Acid Bacteria in Poly(Ethylene Oxide) Nanofibers [J]. Pharmaceutics, 2019, 11(9).

Chandrashekhar P, Minooei F, Arreguin W, et al. Perspectives on Existing and Novel Alternative Intravaginal Probiotic Delivery Methods in the Context of Bacterial Vaginosis Infection [J]. AAPS J, 2021, 23(3): 66.

Han Y, Liu Z, Chen T. Role of Vaginal Microbiota Dysbiosis in Gynecological Diseases and the Potential Interventions [J]. Front Microbiol, 2021, 12: 643422.

Junca H, Pieper D H, Medina E. The emerging potential of microbiome transplantation on human health interventions [J]. Comput Struct Biotechnol J, 2022, 20: 615–27.

Cohen C W, Fontaine K R, Arend R C, et al. A Ketogenic Diet Reduces Central Obesity and Serum Insulin in Women with Ovarian or Endometrial Cancer [J]. J Nutr, 2018, 148(8): 1253–60.

Wallace T C, Bultman S, D'adamo C, et al. Personalized Nutrition in Disrupting Cancer - Proceedings From the 2017 American College of Nutrition Annual Meeting [J]. J Am Coll Nutr, 2019, 38(1): 1–14.

Playdon M C, Nagle C M, Ibiebele T I, et al. Pre-diagnosis diet and survival after a diagnosis of ovarian cancer [J]. Br J Cancer, 2017, 116(12): 1627–37.

Lin C, Zeng Z, Lin Y, et al. Naringenin suppresses epithelial ovarian cancer by inhibiting proliferation and modulating gut microbiota [J]. Phytomedicine, 2022, 106: 154401.

# The Impact of Mitochondria in Ovarian Cancer Cell Metabolism, Proliferation, and Metastasis

7

Heide Schatten

**Abstract**

Mitochondrial dysfunctions are significantly implicated in cancer initiation, progression, and metastasis, which have been shown for several cancers including ovarian cancer.

An increase in mitochondrial dysfunction is also associated with drug resistance along with cancer progression, which in part is related to its specific microenvironment that is characterized by ascites, low glucose levels, and hypoxia that causes ovarian cancer cells to switch to mitochondrial respiration to enable their survival. Peritoneal ascitic fluid accumulation is a specific feature of ovarian cancer, and it is a major cause of its metastatic spread that also presents challenges for effective treatment. Among the treatment difficulties for ovarian cancer is the mutation rate and frequency of mtDNA in ovarian cancer tissue that can affect the efficiency of chemotherapeutic drugs. The varied and multiple mutations of different types enable metabolic reprogramming, cancer cell proliferation, and drug resistance.

New specific information on mechanisms underlying several of the mitochondrial dysfunctions has led to proposing various mitochondrial determinants as targets for ovarian cancer therapy, which include targeting specific mitochondrial proteins and phosphoproteins as well as reactive oxygen species (ROS) that accumulate abnormally in cancer cells. Because of the genetically and histologically heterogeneous nature of the disease, combination therapy approaches will be necessary to combat the disease and achieve progress in effective treatment of ovarian cancer. This chapter will address (1) mitochondrial vulnerabilities underlying dysfunction and disease; (2) mitochondrial dysfunction in ovarian cancer; (3) present treatment difficulties for ovarian cancer and new potential treatment strategies to target ovarian cancer mitochondrial metabolism; and (4) biobehavioral factors influencing ovarian cancer development.

**Keywords**

Mitochondria · Mitochondrial metabolism · Mitochondrial dysfunctions · Warburg effect · Ovarian cancer pathogenesis · Ovarian cancer spheroids · Ovarian cancer therapy · Mitochondria as targets for ovarian cancer treatment · Ovarian cancer cell proliferation · Mitochondrial mutations · Ovarian cancer drug resistance · Biobehavioral factors in ovarian cancer development

H. Schatten (✉)
Department of Veterinary Pathobiology, University of Missouri-Columbia, Columbia, MO, USA
e-mail: SchattenH@missouri.edu



H. Schatten (ed.), *Cell and Molecular Biology of Ovarian Cancer*, Advances in Experimental Medicine and Biology 1452, https://doi.org/10.1007/978-3-031-58311-7_7

## 7.1 Introduction

Mitochondria are indispensable for nearly every eukaryotic organism and play essential roles in multiple cellular functions including energy metabolism with 90% of a cell's ATP being generated by mitochondria, oxidative stress, calcium regulation, metabolic and other cellular signaling, cellular differentiation, apoptosis, cell cycle progression, and cell growth.

Having their own DNA, mitochondrial DNA (mtDNA) differs significantly from nuclear DNA (nDNA) in that it is circular and contains a double-stranded smaller genome of 16.5 kb that encodes 37 proteins, with 13 protein-coding genes being involved in oxidative phosphorylation (OXPHOS) complexes, forming the respiratory chain complexes, 22 tRNAs, and 2 rRNAs (Sharma and Sampath 2019). The majority of proteins of the respiratory chain complexes are encoded by nuclear genes with many of nuclear-encoded proteins being transported into mitochondria (Sharma and Sampath 2019). The difference between mtDNA and nDNA further includes the existence of multiple copies per cell, which can vary depending on energy demand.

Mitochondrial biogenesis is regulated by nDNA-encoded mtDNA genes and by the interplay between the two genomes; nuclear-mitochondrial DNA crosstalks can influence disease processes that play a considerable role in cancer (Choudhury and Singh 2017). Altered energy metabolism is a characteristic feature of cancer which had first been recognized by Otto Warburg (Warburg 1925; Warburg et al. 1924, 1927; reviewed in Liberti and Locasale 2016; Potter et al. 2016) in pioneering studies that built the foundation for subsequent in-depth experimentation on cell, genetic, proteomic, and molecular levels (Li et al. 2018; reviewed in Shukla and Singh 2021; Shen and Zhan 2022; Grieco et al. 2021; Bindra et al. 2021), shedding new light on its role in cancer including ovarian cancer.

By studying rat tumor cells, Warburg's pioneering work determined how these cancer cells fueled their growth by consuming enormous amounts of glucose compared to normal cells, which they break down without oxygen through fermentation. These findings were subsequently well known as the Warburg effect as a most significant feature and hallmark of cancer cells. Based on his pioneering early studies, Warburg proposed in his subsequent studies that dysfunctional mitochondria are the root of aerobic glycolysis (Warburg 1956) and that this is the primary cause of cancer. Warburg's discoveries showing that cellular metabolism is different in cancer cells compared to their noncancerous counterparts have been a cornerstone of modern mitochondrial research in cancer and noncancer cells, and significant new knowledge has been added to the original findings. Recently, new methodologies have allowed us to gain detailed new insights on proteomic, genetic, and phenotypical levels (reviewed in Sharma and Sampath 2019; Shukla and Singh 2021; Shen and Zhan 2022; Yin and Shen 2022), which includes electron and super-resolution microscopy (Grieco et al. 2021) quantification of mtDNA, assays to determine mitochondrial functions and dysfunctions (reviewed in Yin and Shen 2022) including quantitative analysis of the mitochondrial proteome, signaling pathway networks, mass spectrometry-based mitochondrial proteomics, and others.

### 7.1.1 Mitochondrial Vulnerabilities Underlying Dysfunction and Disease

We now also know that compared to nDNA, mtDNA is significantly more vulnerable to mutations that play a major role in various diseases and has clearly been shown in cancer studies including ovarian cancer. The vulnerability to mutations is in a large part related to the lack of histones, inefficient DNA repair mechanism, and the generation of reactive oxygen species (ROS) related to compromised mitochondrial activities (Choudhury and Singh 2017). Compromised mitochondrial functions also affect cellular differentiation, cell signaling, apoptosis, cell cycle, and cell growth (Choudhury and Singh 2017), which all are linked to tumorigenesis, implicating mitochondrial dysfunctions significantly in cancer initiation, progression, and metastasis. The genome of dysfunctional mitochondria can activate the mitochondrial retrograde signaling, controlling the expression of nuclear genes, thereby altering the nuclear gene expression profile in cancer cells.

Numerous studies have revealed more important information on mitochondrial dysfunction as a hallmark of cancers including ovarian cancer (Shen and Zhan 2022; Choudhury and Singh 2017; Signorile et al. 2019), which has led to proposing various mitochondrial determinants as targets for ovarian cancer therapy (reviewed in Shen and Zhan 2022). Other studies have determined progression-mediated changes in mitochondrial morphology promoting adaptation to hypoxic peritoneal conditions in serous ovarian cancer (Grieco et al. 2021) and mitochondrial abnormal phenotypes in epithelial ovarian carcinoma associated with biobehavioral factors (Bindra et al. 2021).

### 7.1.2 Mitochondrial Dysfunction in Ovarian Cancer

The above-mentioned observations are generally true for all ovarian cancer phenotypes that all display complex histopathological characteristics including the majority of ovarian cancers including epithelial-derived tumors, with serous (SER), mucinous, endometrioid (END), clear cell (CC), transitional, squamous, mixed, and undifferentiated types (Bell 2005).

Because of the difficulties to detect ovarian cancer in early stages in part because of the lack of reliable biomarkers and other diagnostic means, ovarian cancer remains the deadliest gynecological cancer in women with a survival rate of less than 30% when the cancer has already spread throughout the peritoneal cavity. It is the fifth leading cause of cancer-related deaths in women but early detection increases the survival of affected women to 92% (The American Cancer Society 2020). This genetically and histologically heterogeneous disease is thought to originate in the epithelial cells of the fimbriae of the fallopian tubes. It is mostly spread throughout the peritoneal cavity, and exfoliation from the original tumor through transport by the ascites fluid results in metastases at the first metastatic site, the omentum, which occurs within hours after exfoliation (Lengyel 2010).

While the pathogenesis of ovarian cancer is not yet fully known, several causal factors have been implicated and include genetics, ovulation, environmental factors, and hormones (Nowak and Klink 2020).

New research to determine the exact causes for mitochondrial dysfunctions in ovarian cancer is still actively being pursued, and new therapies to target specific mitochondrial dysfunctions are much needed to achieve progress on ovarian cancer control, healing, and eradication.

As major mitochondrial dysfunctions are related to ROS accumulations which is strongly associated with aging (reviewed in Chistiakov et al. 2014), there is increasing evidence that mitochondrial dysfunction drives ovarian aging (Chiang et al. 2020) and associated vulnerabilities to develop ovarian cancer. We also know that mtDNA content and copy numbers decline in the aging ovary.

An increase in mitochondrial dysfunction is associated with cancer progression and drug resistance (reviewed in Shen and Zhan 2022) which in part is related to its specific microenvironment. Ovarian cancer develops as solid tumor in the abdominal cavity and the ovarian cancer microenvironment is characterized by ascites, low glucose levels, and hypoxia, which causes ovarian cancer cells to switch to mitochondrial respiration to enable their survival. Ovarian cancer is distinguished from other cancers in that peritoneal ascitic fluid accumulation is a specific feature of ovarian cancer and it is a major cause of its metastatic spread, which also presents difficulties for effective treatment.

As ovarian tumors grow bigger, cells typically form dense spherical clusters termed spheroids that are shed into the ascitic fluid, which subsequently can invade surface tissues to form new tumors. These characteristics of ovarian cancer add to the treatment difficulties with chemotherapies such as paclitaxel and cisplatin/carboplatin (Emmings et al. 2019). Spheroids consist of an outer division-prone cellular layer that can survive and proliferate using limited amounts of glucose and oxygen as is available in the ascites. While the outer layer cells of the spheroids can be eliminated with targeted chemotherapy, cells inside of spheroids are not responsive to general chemotherapy treatment, as they densely populate the area that is hypoxic and poorly vascularized, causing these cells to become quiescent. All spheroid cells depend on OXPHOS for ATP production and survival. While the spheroid's proliferating cancer cells are localized in the cancer microenvironment that provides adequate levels of oxygen and glucose but require high levels of ATP for their proliferation, the quiescent cancer cells are localized in microenvironments with low glucose and hypoxia and do not require high levels of ATP as they are quiescent. These cells need different treatments for their elimination, which again presents difficulties finding adequate treatment strategies for these different cell types. In addition, ovarian cancer cells typically contain mitochondria with different mutations, which adds different factors for treatment considerations (Emmings et al. 2019). New strategies have been proposed to include targeting specific mitochondrial proteins and phosphoproteins in ovarian cancers as potentially effective diagnostic and prognostic markers that might be druggable therapeutic targets for ovarian cancer (reviewed in Shukla and Singh 2021). The review by Shukla and Singh (2021) provides excellent details on mitochondrial genetic aspects including mtDNA content in ovarian cancer; mtDNA gene polymorphisms, gene expression, and ovarian cancer risk; gene encoding regions of mtDNA; nuclear DNA-encoded mitochondrial biogenesis genes; expression of energy metabolism genes; and mtDNA alterations and mechanism associated with ovarian cancer chemoresistance that are not addressed in this chapter.

### 7.1.3 Present Treatment Difficulties for Ovarian Cancer and New Potential Treatment Strategies to Target Ovarian Cancer Mitochondrial Metabolism

The present treatments for ovarian cancer mainly consist of cytoreductive surgery combined with cisplatin/paclitaxel (Guerra et al. 2017; reviewed in Shukla and Singh 2021) chemotherapy. The initial response rate to these first-line treatments is high, but most frequently relapse occurs in most ovarian cancer patients with a poor outcome of a 5-year survival rate of only up to 46%, which in part is related to a late diagnosis and a high incidence of chemoresistance (Lheureux et al. 2019). Improvement of the early-stage diagnostic rate of ovarian cancer will be important to achieve more effective treatment and higher success rates for ovarian cancer control (Samuel and Carter 2017).

As the number of mitochondrial DNAs (mtDNAs) is significantly increased in ovarian cancers relative to controls (Wang et al. 2006) and as mitochondrial dysfunctions have clearly been implicated in ovarian cancers (Li et al. 2020; Li and Zhan 2019, 2020) mitochondria are thought to present a most suitable target for developing new chemotherapy approaches.

Several molecular targeted drugs have been developed against mitochondrial dysfunction pathways in ovarian cancer, including olive leaf extract, nilotinib, salinomycin, *Sambucus nigra* agglutinin, tigecycline, and eupatilin, which is aimed at affecting abnormal amino acid metabolism, fatty acid metabolism, and glucose metabolism that widely occur during tumorigenesis (Finley 2019). However, many potential targets have not yet been pursued in molecular detail.

Among the treatment difficulties for ovarian cancer is the mutation rate and frequency of mtDNA in ovarian cancer tissue that can affect the efficiency of chemotherapeutic drugs, as mitochondria play a decisive role in apoptosis and as chemotherapeutic drugs generally eliminate tumor cells by inducing apoptosis. The extent of mitochondrial resistance to apoptosis can vary depending on the quantity and different types of mitochondrial mutations, which will likely affect responses to drug treatments. mtDNA depletion favors pro-survival and anti-apoptotic pathways that may lead to increases in cancer cell proliferation in addition to displaying a chemoresistant phenotype (Guerra et al. 2017).

The varied and multiple mutations of different types can lead to mitochondrial dysfunction resulting in metabolic reprogramming, cancer cell proliferation, and drug resistance, as shown for human ovarian cancer cells in culture. In these cultured cells, depletion of mtDNA resulted in mitochondrial dysfunction leading to chemoresistance, and the cells displayed a significant shift toward a more oxidative metabolism. Such experiments clearly show how important mitochondria are for the altered cell metabolism in cancer cells and how they are influenced by numerous environmental conditions (Guerra et al. 2017).

Ovarian drug-resistant cells display morphological and structural differences with mitochondrial network reorganization including defects in the electron transport chain/mitochondrial coupling as shown for cisplatin-resistant cells compared to cisplatin-sensitive cells (Guerra et al. 2017; reviewed in Shukla and Singh 2021).

These findings have led to proposing new mitochondrial therapies to target glycolysis, glutamine metabolism, mitochondrial retrograde signaling, mitochondrial OXPHOS complexes, and mutant mtDNA (reviewed in Shukla and Singh 2021). For example, the glycolysis inhibitor lonidamine in combination with the drug doxorubicin yielded better therapeutic efficacy for ovarian cancer as well as other cancer treatments. The mitochondrial enzyme, glutaminase, in combination with mTOR inhibitors sensitizes ovarian cancer cells to paclitaxel (Guerra et al. 2017; reviewed in Shukla and Singh 2021). Furthermore, electron transport chain inhibitors including rotenone, oligomycin A, and metformin as well as mitochondrial uncouplers including valinomycin and dinitrophenol were shown to be antineoplastic. Also, the OXPHOS inhibitor, VLX600, reduced the viability of quiescent cells to cause spheroid reduction and thereby inhibition of tumor cell proliferation (Guerra et al. 2017; reviewed in Shukla and Singh 2021). Other studies have shown that doxycycline and salinomycin inhibit the proliferation of ovarian cancer cells and selectively eliminate cisplatin-resistant ovarian cancer cells. The pyruvate dehydrogenase activator, CPI-613, inhibits mitochondrial ATP production, leading to ovarian cancer cell death (Guerra et al. 2017; reviewed in Shukla and Singh 2021). The metformin, complex I inhibitor, inhibits tumor cell proliferation, metastasis, and angiogenesis which was further enhanced when it was combined with cisplatin/paclitaxel (Guerra et al. 2017; reviewed in Shukla and Singh 2021).

These studies show the high potential of targeting specific aspects of mitochondrial metabolism and dysfunctions as still wide-open possibilities for treatment of ovarian cancer.

### 7.1.4 Biobehavioral Factors Influencing Ovarian Cancer Development

The importance of healthy mitochondria for cancer prevention and therapy has also been shown in studies with patients to probe biobehavioral factors (Bindra et al. 2021). In these studies, multiple biochemical and molecular markers of mitochondrial content and function were quantified in tissue from 51 patients with benign ovarian masses and 128 patients with high-grade epithelial ovarian cancer. The results showed that human tumor cell mitochondria play a pivotal role in the growth of malignant tumors and that mitochondrial function enables tumor cell adaptation to stressful environments, providing plasticity for tumor growth and survival, conferring resistance to apoptotic signaling, and promoting the metastatic phenotypes (Wallace 2012; Sotgia et al. 2012; Vyas et al. 2016). These studies further showed that human tumor cells exhibit altered energy metabolism and metabolic reprogramming associated with alterations in mitochondrial DNA (mtDNA) and its associated proteins (Weinberg and Chandel 2015) and that ovarian tumor cells secrete pro-inflammatory mediators like interleukin-6 (IL-6), which in turn promotes tumor proliferation, invasion, and chemoresistance (Macciò and Madeddu 2012).

As mitochondria have been shown to dynamically respond to biobehavioral processes, glucocorticoid stress hormones released during psychosocial stress might affect cancer growth as well as treatments. With these studies in mind (Bindra et al. 2021), perhaps psychosomatic treatment strategies may also be considered in combination with other therapies including surgery and chemotherapy, as mitochondria respond to biobehavioral aspects (Bindra et al. 2021).

## 7.2 Conclusions and Future Perspectives

New research to determine the exact causes for mitochondrial dysfunctions in ovarian cancer is still actively being pursued, and new therapies to target specific mitochondrial dysfunctions are much needed to achieve progress on ovarian cancer control, healing, and eradication.

The present treatments for ovarian cancer mainly consist of cytoreductive surgery combined with cisplatin/paclitaxel chemotherapy. The initial response rate to these first-line treatments is high but most relapse occurs in most ovarian cancer patients with a poor outcome of a 5-year survival rate of only up to 46%, which in part is related to a late diagnosis and a high incidence of chemoresistance. Improvement of the early-stage diagnostic rate of ovarian cancer will be important to achieve more effective treatment and a higher success rate for ovarian cancer control.

These findings have led to proposing new mitochondrial therapies to target glycolysis, glutamine metabolism, mitochondrial retrograde signaling, mitochondrial OXPHOS complexes, and mutant mtDNA, and new strategies have been proposed to include targeting specific mitochondrial proteins and phosphoproteins in ovarian cancers as potentially effective diagnostic and prognostic markers that might be druggable therapeutic targets for ovarian cancer, but many potential targets have not yet been pursued in molecular detail.

There is still high potential for targeting specific aspects of mitochondrial metabolism and dysfunctions to pursue still wide-open possibilities for treatment of ovarian cancer.

## References

Bell DA (2005). Origins and molecular pathology of ovarian cancer. Mod. Pathol., 18 (Suppl 2), S19–S32

Bindra S, McGill MA, Triplett MK, Tyagi A, Thaker PH, Dahmoush L, Goodheart MJ, Ogden RT, Owusu-Ansah E, R Karan K, Cole S, Sood AK, Lutgendorf SK, Picard M (2021). Mitochondria in epithelial ovarian carcinoma exhibit abnormal phenotypes and blunted associations with biobehavioral factors. Sci Rep 11(1):11595. https://doi.org/10.1038/s41598-021-89934-6. PMID: 34078919; PMCID: PMC8172869.

Chiang JL, Shukla P, Pagidas K, Ahmed NS, Karri S, Gunn DD, Hurd WW, Singh KK (2020). Mitochondria in ovarian aging and reproductive longevity. Ageing Res Rev 63:101168.

Chistiakov DA, Sobenin IA, Revin VV, Orekhov AN, Bobryshev YV (2014). Mitochondrial aging and age-related dysfunction of mitochondria. Biomed Res Int. 238463. https://doi.org/10.1155/2014/238463. PMID: 24818134; PMCID: PMC4003832.

Choudhury AR, Singh KK. Mitochondrial determinants of cancer health disparities. Semin Cancer Biol. 2017 Dec;47:125–146. https://doi.org/10.1016/j.semcancer.2017.05.001. Epub 2017 May 6. PMID: 28487205; PMCID: PMC5673596.

Emmings E, Mullany S, Chang Z, Landen CN Jr, Linder S, Bazzaro M (2019). Targeting Mitochondria for Treatment of Chemoresistant Ovarian Cancer. Int J Mol Sci 20(1):229. https://doi.org/10.3390/ijms20010229. PMID: 30626133; PMCID: PMC6337358.

Finley LWS (2019) Metabolic signal curbs cancer-cell migration. Nature, vol. 571, no. 7763, pp. 39–40.

Grieco JP, Allen ME, Perry JB, Wang Y, Song Y, Rohani A, Compton SLE, Smyth JW, Swami NS, Brown DA, Schmelz EM (2021). Progression-Mediated Changes in Mitochondrial Morphology Promotes Adaptation to Hypoxic Peritoneal Conditions in Serous Ovarian Cancer. Front Oncol 10:600113. https://doi.org/10.3389/fonc.2020.600113. PMID: 33520711; PMCID: PMC7838066.

Guerra F, Arbini AA, Moro L (2017). Mitochondria and cancer chemoresistance. Biochim Biophys Acta Bioenerg 1858(8):686–699. https://doi.org/10.1016/j.bbabio.2017.01.012. PMID: 28161329.

Lengyel E (2010). Ovarian cancer development and metastasis. Am J Pathol 177(3):1053–64. https://doi.org/10.2353/ajpath.2010.100105

Lheureux S, Gourley C, Vergote I, Oza AM (2019). Epithelial ovarian cancer. Lancet

393(10177):1240–1253. https://doi.org/10.1016/S0140-6736(18)32552-2. PMID: 30910306.

Li N, Li H, Cao L, Zhan X (2018). Quantitative analysis of the mitochondrial proteome in human ovarian carcinomas. Endocr Relat Cancer 25(10):909–931. https://doi.org/10.1530/ERC-18-0243. PMID: 29997262.

Li N, Zhan X (2019). Signaling pathway network alterations in human ovarian cancers identified with quantitative mitochondrial proteomics. EPMA J 10(2):153–172. https://doi.org/10.1007/s13167-019-00170-5. PMID: 31258820; PMCID: PMC6562010.

Li N, Zhan X (2020). Mass spectrometry-based mitochondrial proteomics in human ovarian cancers. Mass Spectrom Rev 39(5-6):471–498. https://doi.org/10.1002/mas.21618. PMID: 32020673.

Li N, Li H, Wang Y, Cao L, Zhan X (2020). Quantitative proteomics revealed energy metabolism pathway alterations in human epithelial ovarian carcinoma and their regulation by the antiparasite drug ivermectin: data interpretation in the context of 3P medicine. EPMA J 11(4):661–694. https://doi.org/10.1007/s13167-020-00224-z. PMID: 33240452; PMCID: PMC7680500.

Liberti MV, Locasale JW (2016). The Warburg Effect: How Does it Benefit Cancer Cells? Trends Biochem Sci 41(3):211–218. https://doi.org/10.1016/j.tibs.2015.12.001. Erratum in: Trends Biochem Sci. 2016 Mar;41(3):287. Erratum in: Trends Biochem Sci. 2016 Mar;41(3):287. PMID: 26778478; PMCID: PMC4783224.

Macciò A, Madeddu C (2012). Inflammation and ovarian cancer. Cytokine 58(2):133–47. https://doi.org/10.1016/j.cyto.2012.01.015. PMID: 22349527.

Nowak M, Klink M (2020). The role of tumor-associated macrophages in the progression and chemoresistance of ovarian cancer. Cells. 22;9(5):1299.

Potter M, Newport E, Morten KJ (2016). The Warburg effect: 80 years on. Biochem Soc Trans 44(5):1499–1505. https://doi.org/10.1042/BST20160094. PMID: 27911732; PMCID: PMC5095922.

Samuel P, Carter DR (2017). The Diagnostic and Prognostic Potential of microRNAs in Epithelial Ovarian Carcinoma. Mol Diagn Ther 21(1):59–73. https://doi.org/10.1007/s40291-016-0242-z. PMID: 27718164.

Sharma P, Sampath H (2019). Mitochondrial DNA Integrity: Role in Health and Disease. Cells 8(2):100. https://doi.org/10.3390/cells8020100. PMID: 30700008; PMCID: PMC6406942.

Shen L, Zhan X (2022). Mitochondrial Dysfunction Pathway Alterations Offer Potential Biomarkers and Therapeutic Targets for Ovarian Cancer. Oxid Med Cell Longev 2022:5634724. https://doi.org/10.1155/2022/5634724. PMID: 35498135; PMCID: PMC9045977.

Shukla P, Singh KK (2021). The mitochondrial landscape of ovarian cancer: emerging insights. Carcinogenesis 42(5):663–671. https://doi.org/10.1093/carcin/bgab033. PMID: 33928357; PMCID: PMC8163040.

Signorile A, De Rasmo D, Cormio A, Musicco C, Rossi R, Fortarezza F, Palese LL, Loizzi V, Resta L, Scillitani G, Cicinelli E, Simonetti F, Ferretta A, Russo S, Tufaro A, Cormio G (2019). Human Ovarian Cancer Tissue Exhibits Increase of Mitochondrial Biogenesis and Cristae Remodeling. Cancers (Basel) 11(9):1350. https://doi.org/10.3390/cancers11091350. PMID: 31547300; PMCID: PMC6770021.

Sotgia F, Whitaker-Menezes D, Martinez-Outschoorn UE, Salem AF, Tsirigos A, Lamb R, Sneddon S, Hulit J, Howell A, Lisanti MP (2012). Mitochondria "fuel" breast cancer metabolism: fifteen markers of mitochondrial biogenesis label epithelial cancer cells, but are excluded from adjacent stromal cells. Cell Cycle 11(23):4390–401. https://doi.org/10.4161/cc.22777. PMID: 23172368; PMCID: PMC3552922.

The American Cancer Society. Cancer Facts and Figures. (2020). Available at: https://www.cancer.org/cancer/ovarian-cancer.html.

Vyas S, Zaganjor E, Haigis MC (2016). Mitochondria and Cancer. Cell 166(3):555–566. https://doi.org/10.1016/j.cell.2016.07.002. PMID: 27471965; PMCID: PMC5036969.

Wallace DC (2012). Mitochondria and cancer. Nat Rev Cancer 12:685–698.

Wang Y, Liu VW, Xue WC, Cheung AN, Ngan HY (2006). Association of decreased mitochondrial DNA content with ovarian cancer progression. Br J Cancer 95(8):1087–91. https://doi.org/10.1038/sj.bjc.6603377. PMID: 17047655; PMCID: PMC2360719.

Warburg O (1925). The metabolism of carcinoma cells. The Journal of Cancer Research 9(1):148–163.

Warburg O, Posener K, Negelein E (1924). Ueber den Stoffwechsel der Tumoren. Biochemische Zeitschrift 152(1):319–344.

Warburg O, Wind F, Negelein E (1927). The metabolism of tumors in the body. The Journal of general physiology 8(6):519–530. [PubMed: 19872213]

Warburg O (1956). On the origin of cancer cells. Science 123(3191):309–314.

Weinberg SE, Chandel NS (2015). Targeting mitochondria metabolism for cancer therapy. Nat Chem Biol 11(1):9–15. https://doi.org/10.1038/nchembio.1712. PMID: 25517383; PMCID: PMC4340667.

Yin Y, Shen H (2022). Common methods in mitochondrial research (Review). Int J Mol Med 50(4):126. https://doi.org/10.3892/ijmm.2022.5182. PMID: 36004457; PMCID: PMC9448300.

# Index


**A**
ABC transporter, 12, 72
Acetylation, 26–28
Apoptosis evasion, 12, 25, 76–77
Autophagy, 23, 74–75

**B**
Bacteria, 10, 45, 98, 100–103, 108–115
Biobehavioral factors in ovarian cancer development, 121, 123, 124

**C**
Cancer stem cells, 77
Cell cycle, 2, 3, 38–43, 46–49, 51, 52, 71, 81, 120
Centrioles, 4, 27, 39, 41–46, 48, 51, 52
Centrosomal proteins, 41–48
Centrosome clustering, 42, 49–50, 54–56
Centrosome phosphorylation, 46
Centrosomes, vi, 3, 4, 38–56
Checkpoint inhibitors, 81

**D**
Detyrosination, 26, 27
Disease classification, 66, 69
Disease relapse, 7, 75, 78, 80
Docetaxel, 9, 73
Drug resistance, 12, 49, 50, 54, 56, 71–78, 112, 121, 123
Dynamic instability, 2, 3, 23
Dysbiosis, 98, 100–103, 109, 111, 114

**E**
End-binding protein 1 (EB1), 24, 25
Epithelial ovarian cancer (EOC), 6, 7, 9, 11, 38, 45, 48, 52–54, 66–89, 98–103, 107–109, 113, 114, 123

**G**
Gastrointestinal tract, 101–102
Gut microbes, 101, 109, 112
Gut microbiota, 107–115

**H**
Human microbiome, 98–102

**I**
Immune regulation, 110, 111
Immunotherapies, 53, 54, 83, 88, 112, 113
Ixabepilone, 10

**L**
Lower reproductive tract, 99–101

**M**
Metabolism, 27, 39, 73, 76, 85, 98, 100, 110, 113, 120–124
Microbiota, 100–102, 107–115
Microorganisms, 98, 99, 109, 111, 114, 115
Microtentacle, 29, 30
Microtubule, 1, 22, 38, 70
Microtubule-associated proteins (MAPs), 2, 3, 9, 12, 22–27, 44, 50, 52, 66, 67
Microtubule organization, 38, 40, 44, 45, 47
Mitochondria, 5, 39, 40, 120–124
Mitochondria as targets for ovarian cancer treatment, 120–124
Mitochondrial dysfunctions, 121–124
Mitochondrial metabolism, 122–124
Mitochondrial mutations, 123
mRNAs, 22, 25, 74, 77, 81, 87
Multipolar spindles, 39, 48



H. Schatten (ed.), *Cell and Molecular Biology of Ovarian Cancer*, Advances in Experimental Medicine and Biology 1452, https://doi.org/10.1007/978-3-031-58311-7

**O**
Oncogenic pathways, 49
Ovarian cancer, 5, 21, 38, 66, 97, 107, 120
Ovarian cancer cell proliferation, 49, 55, 110
Ovarian cancer development, 100, 110, 123–124
Ovarian cancer drug resistance, 49, 50
Ovarian cancer pathogenesis, 38, 110, 115
Ovarian cancer spheroids, 121–123
Ovarian cancer therapy, 1–14, 38–56, 84, 112, 121

**P**
Paclitaxel, 6–12, 14, 22–26, 49, 54, 55, 70, 71, 79, 81–83, 85, 112, 122–124
Personalized cancer therapies, 53, 56, 113
Polyglutamylation, 26, 27, 44
Polyglycylation, 26, 28
Post-translational modifications, 2
Primary cilia, 28, 41, 43, 45, 50–54

**R**
Reproductive tract microbiota, 107–109
Resistance pathways, 73–74

**T**
Tau, 9, 12, 22, 23
Therapeutic targets, 22, 38–56, 71, 79, 81, 122, 124
Tubulin, 1–6, 10, 12, 13, 21–22, 24–29, 44, 71
Tubulin code, 2, 26, 29
Tumor microbiota, 108
Tumor suppressor genes, 40, 43, 48, 56

**U**
Upper reproductive tract, 99–101, 109

**V**
Vaccines, 86–88, 113
Vaginal microbes, 99
Vinblastine, 11, 23, 26
Vincristine, 10–12, 23

**W**
Warburg effect, 76, 120

The manufacturer's authorised representative in the EU is Springer Nature Customer Service Centre GmbH, Europaplatz 3, 69115 Heidelberg, Germany. If you have any concerns regarding our products, please contact ProductSafety@springernature.com

Printed and bound by CPI Group (UK) Ltd, Croydon, CR0 4YY
30/09/2026
02225577-0019